Common Envelope Evolution

AMERICAN ASTRONOMICAL SOCIETY IOP | ebooks™

About the program:

AAS-IOP Astronomy ebooks is the official book program of the American Astronomical Society (AAS), and aims to share in depth the most fascinating areas of astronomy, astrophysics, solar physics and planetary science. The program includes publications in the following topics:

Books in the program range in level from short introductory texts on fast-moving areas, graduate and upper-level undergraduate textbooks, research monographs and practical handbooks.

For a complete list of published and forthcoming titles, please visit iopscience.org/books/aas.

About the American Astronomical Society

The American Astronomical Society (aas.org), established 1899, is the major organization of professional astronomers in North America. The membership (~7,000) also includes physicists, mathematicians, geologists, engineers and others whose research interests lie within the broad spectrum of subjects now comprising the contemporary astronomical sciences. The mission of the Society is to enhance and share humanity's scientific understanding of the universe.

Editorial Advisory Board

Common Envelope Evolution

Natalia Ivanova
University of Alberta

Stephen Justham
University of Amsterdam and University of the Chinese Academy of Sciences

Paul Ricker
University of Illinois

IOP Publishing, Bristol, UK

ISBN 978-0-7503-1563-0 (ebook)
ISBN 978-0-7503-1561-6 (print)
ISBN 978-0-7503-1909-6 (myPrint)
ISBN 978-0-7503-1562-3 (mobi)

DOI 10.1088/2514-3433/abb6f0

Version: 20201201

AAS–IOP Astronomy
ISSN 2514-3433 (online)
ISSN 2515-141X (print)

British Library Cataloguing-in-Publication Data: A catalogue record for this book is available from the British Library.

Published by IOP Publishing, wholly owned by The Institute of Physics, London

IOP Publishing, Temple Circus, Temple Way, Bristol, BS1 6HG, UK

US Office: IOP Publishing, Inc., 190 North Independence Mall West, Suite 601, Philadelphia, PA 19106, USA

Contents

Author Biographies

Natalia Ivanova

Natalia (known as Natasha) Ivanova is Professor of theoretical and computational astrophysics at the Physics Department, University of Alberta. Her scientific interests include everything about understanding of single, binary, multiple stars and clusters of them, stellar physics, and numerical codes that can create a star or many of them inside a computer. She focuses mainly on modeling extreme interactions between the stars, such as common-envelope events, collisions and mass transfers. In 2010 she was appointed as Canada Research Chair in Astronomy and Astrophysics. Before that, she was a CITA postdoctoral fellow distinguished with Tremaine Fellowship at the University of Toronto (Canada), and a theoretical postdoctoral fellow at the Northwestern University (USA). She obtained her DPhil from the University of Oxford (UK) and studied as an undergraduate at Saint-Petersburg State University (Russia). In her spare time, she likes to mountaineering, hiking, diving, and is a certified downhill ski instructor. She is married with two children, has a snow-white Labrador retriever, and lives in Edmonton, Canada.

Stephen Justham

Stephen Justham is currently in the University of Amsterdam, where he is now an acting group leader, on extended leave from a Professorship in the University of the Chinese Academy of Sciences. Stephen especially aims to understand the physics and consequences of stellar interactions, including how those help to explain the observed variety of stellar systems and explosive transients. Stephen studied physics as an undergraduate in Cambridge, then stellar astrophysics as a graduate student in the Open University, Milton Keynes, and as a postdoc in the University of Oxford. He moved to Beijing in 2008, first to the Kavli Institute of Astronomy and Astrophysics, then as an international research fellow in the National Astronomical Observatory of the Chinese Academy of Sciences, eventually joining the University of the Chinese Academy of Sciences, in which he was appointed Full Professor in early 2016.

Paul Ricker

Paul M. Ricker is Professor of Astronomy at the University of Illinois. His primary research interests lie in the application of hydrodynamical simulation to galaxy clusters and interacting binary stars. He held postdoctoral appointments at the University of Chicago and University of Virginia, and before that earned his PhD in physics at the University of Chicago and his BS in physics and astronomy at the Pennsylvania State University. He is one of the principal authors of the widely-used Flash simulation code, sharing in the 2000 Gordon Bell Prize, and in 2001 he received the Presidential Early Career Award for Scientists and Engineers (PECASE). His other interests include astrophotography, sea kayaking, and bicycle touring. He is married with three cats and lives in Champaign.

Foreword

This book is intended as a graduate-level textbook, and it generally assumes prior knowledge of stellar evolution and stellar physics at the level of an introductory graduate course.

Classic introductions to stellar evolution include the textbooks by Schwarzschild (1958; "Structure and Evolution of the Stars"), Hansen, Kawaler, and Trimble (2004; "Stellar Interiors"), and Kippenhahn, Weigert, and Weiss (2012; "Stellar Structure and Evolution"). Books that also present the fundamentals of interacting binary-star evolution include Pringle and Wade (1985; "Interacting Binary Stars"), Eggleton (2006; "Evolutionary Processes in Binary and Multiple Stars"), and Eldridge and Tout (2019; "The Structure and Evolution of Stars").

The topic is a rapidly-moving field, and many literature references are provided, but the text is intended to explain the physics and scientific background, not cover the full scope of the literature. This book represents the authors' perspective only, and it should not be taken as an exhaustive review.

As evidence of the fact that the field is vibrant, the authors enjoyed vigorous internal debates in preparing this book. We think we learned from those discussions, and we hope the results are enlightening to the readers as well as the authors. Much of this book was written during the interesting times of COVID-19, with animated discussion among the authors over numerous hours on video calls.

NI is extremely grateful to her D.Phil. adviser Philipp Podsiadlowski, who inspired her two decades ago to study common envelope systems and was a unique mentor. NI also thanks Ron Taam, Fred Rasio, and Vicky Kalogera for their invaluable introduction in a broad field of interacting binaries and a plethora of interesting puzzles in physics one can find there. NI thanks her family for support during the unusual pandemic-style time the book was getting written.

SJ adds his own thanks to Philipp Podsiadlowski for introducing him to this and other fields, and for insight, thoughtful mentorship, and considerable patience over many years. He is grateful to all the people who have provided him with kindness, explanations, and stimulating discussions. SJ also thanks Selma de Mink, and her stellar group, for providing very interesting, energetic, and enjoyable scientific hospitality in Amsterdam.

PR thanks especially Ron Taam, who first introduced him to common-envelope evolution; You-Hua Chu, who first raised the planetary nebula connection with him; and Ron Webbink, who has been a font of knowledge on binary interaction and mass transfer. Each has been a valued mentor for many years. PR is grateful to his wife, Kathleen, and his students for putting up with the time spent writing this book.

Much of the early development of this book occurred during the KITP STARS17 program, which was supported in part by the National Science Foundation under grant PHY 17-48958. NI also acknowledges support from CRC program and NSERC Discovery. PR also acknowledges support from NSF under grant AST 14-13367.

We also thank Dr Leigh Jenkins, Poppy Emerson, Sarah Armstrong, and everyone at IOP Publishing for their forbearance and support.

The authors collectively thank everyone who provided help in improving the content of this book; any errors are, of course, our own.

Natalia Ivanova, Stephen Justham and Paul Ricker

Chapter 1

Introduction

Common-envelope evolution (CEE) is considered a vital process in the formation of many of the most exciting systems in astrophysics. For example, this mechanism is thought to be instrumental in the formation of stellar-mass gravitational-wave merger sources, because it aids the formation of compact-object binaries which are sufficiently close for gravitational waves to lead to a merger within a Hubble time. The significance of understanding CEE extends to a much wider set of stellar systems. If our present knowledge of stellar evolution is correct, we must understand CEE to explain diverse stellar systems and transients.

Figure 1.1 presents a schematic description of canonical CEE. A binary star consisting of a giant star with a denser companion enters a temporary merger, during which the companion star orbits inside the envelope of the giant. As the companion "inspirals" through the envelope, energy and angular momentum are transferred to the envelope, which may lead to ejection of the envelope. If the envelope of the giant is ejected, the initial wide binary is converted into a close binary. If ejection fails, then the stars merge. This would potentially produce a star with properties that would not be expected from single-star evolution. Note that this means understanding common-envelope evolution is not only important for explaining the formation of many compact binaries, but also for the formation of post-merger single stars.

Significant questions remain about all the stages illustrated in Figure 1.1: which systems *enter* CEE, what happens *during* CEE, and what will be *the final state* of any given CEE. So major uncertainties still exist, about which reasonable scientists disagree, and not only regarding fine details. This book describes our present understanding of CEE and how the community is attempting to make progress. We do not provide an introduction to stellar evolution, or to the theory of the evolution of interacting binary stars. This Chapter gives an overview of why CEE is thought to happen, and why it is considered important, and a brief precise of why modeling it is difficult.

doi:10.1088/2514-3433/abb6f0ch1

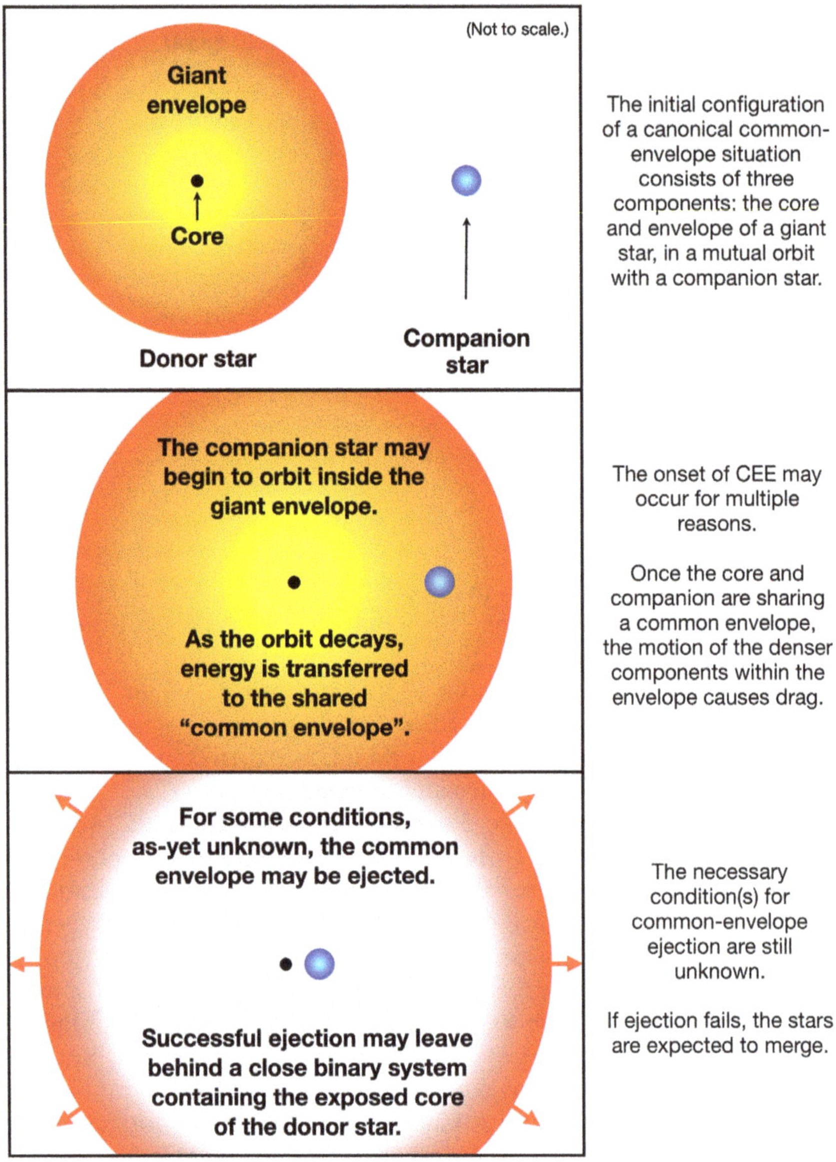

Figure 1.1. Schematic of the overall problem of canonical common-envelope evolution (proceeding top to bottom). The denser component in a binary-star system starts to move inside the envelope of the less dense star. As the orbit decays, orbital energy is transferred to the envelope. If the envelope is successfully ejected, this process may leave behind a binary with a much smaller orbital separation than before the onset of the common-envelope phase. Moreover, one of the components of a surviving post-CE binary is expected to have lost a large fraction of its envelope.

1.1 Why Do We Think Common-envelope Evolution Happens?

Understanding why we infer CEE to be important is a good exercise in astrophysical detective work, combining indirect clues from observed systems with fundamentals from stellar evolution. These clues allow us to conclude that a process exists that can shrink binary orbits by orders of magnitude, producing close binaries with orbital periods as short as minutes or hours. We can additionally infer that this mechanism operates very rapidly compared to stellar-evolution timescales.

The paper most commonly cited for the origin of CEE as an important physical process is actually a conference proceedings article. Paczynski (1976) argues that a particular observed system—V471 Tau—is evidence for a mechanism that can significantly shrink binary orbits, that CEE is the most natural way to explain this, and that this would also explain the formation of a broader class of compact binary stars. We explain this argument below. However, Paczynski (1976) also cites previous work that discussed the possibility of forming a "common-envelope binary," and states that Jeremiah Ostriker (1973, private communication) and Ron Webbink (1975) have previously argued in favor of the common-envelope formation mechanism for those "cataclysmic binaries."

The binary V471 Tau had been discovered in the Hyades open cluster. V471 Tau contains a white dwarf and a low-mass main-sequence star, orbiting each other with a period of about 12 hours. Each of the components was inferred to be about 0.8 $M_{\odot}$. It was a key system because the white dwarf looked particularly young, i.e., it had not spent long cooling after being exposed. Hence the white dwarf was inferred to have been formed from a star that had only recently left the main sequence. The age of the Hyades cluster was known—specifically, the mass of stars which were then leaving the main sequence was about 2 $M_{\odot}$. Therefore, astronomers had a good idea of the mass of the star that made the white dwarf.

The properties of this binary apparently required that its separation had decreased significantly, and rapidly—i.e., in only a small fraction of the evolutionary timescale. Reflecting common wisdom at the time, Paczyński noted that, in order to make an $\approx$0.8 $M_{\odot}$ white dwarf, a $\approx$2 $M_{\odot}$ star needs to climb the asymptotic giant branch (AGB). Such a star was expected to have a radius of about 600 $R_{\odot}$, indicating that the orbital period of the binary at the time would have been expected to be around 10 years or more. Paczyński speculated that it might have been possible to make the period as short as a year, given freedom to vary some details, but not less. So a mechanism was needed to relatively rapidly change the orbital period from at least a year—probably an order of magnitude more than that—to only 12 hours.

The onset of the common-envelope phase is only briefly considered in Paczynski (1976). It is natural to assume that, whatever mechanism shortened the orbital period, it required the two stars to be interacting. If a 2 $M_{\odot}$ AGB star fills its Roche lobe and starts to transfer mass onto a 0.8 $M_{\odot}$ companion, mass transfer was expected to happen on a dynamical timescale, at a very high rate.[1]

[1] The component masses in V471 Tau were known only approximately, but the uncertainty was not large enough to affect the argument Paczyński was making.

Paczyński's original, extremely influential, paper states about the consequences of this mass transfer: "It seems very likely that under such conditions a common envelope is formed around a binary system. The initial phases of this process when the main-sequence dwarf is being enveloped by the red supergiant are difficult to analyze." As modern astrophysicists, trying to be worthy to build on the work of Paczyński, this is reassuring; this phase is indeed "difficult to analyze." It remains one of the more troublesome phases to simulate, with modern simulations typically artificially skipping past the long onset to save computer time (see Chapters 4 and 5).

Paczyński then went on to sketch the inspiral of the core of the giant, together with the low-mass main-sequence star, as they experience drag on their orbits within their common envelope. As the components spiral inwards, they transfer energy and angular momentum to the envelope. Those few paragraphs combine estimates from energetic arguments and angular momentum conservation in plausibility arguments, and describe how they may fit together, but they do not give a definitive picture of what will happen, nor how or when the envelope would be eventually ejected. The remainder of this book will give a modern account of how and when we think common-envelope inspiral and envelope ejection occur.

Paczyński also suggested that, were the envelope to be ejected, the immediate post-common-envelope configuration would appear as a planetary nebula with a close binary in the center. We now have multiple instances of such systems, supporting his predictions in at least those cases, and acting as further clear evidence that CEE does really occur. Chapter 10 will go into more detail about the present state of comparison between specific observational examples and models of CEE.

1.2 Why Is Common-envelope Evolution Broadly Important?

A simple answer to this question is that "bringing stars closer together increases the chance that they transfer mass." A more subtle explanation is that CEE enables close binaries to exist for which one, or both, of the components needed to have evolved in a much wider binary than they are currently observed to be. Stars, broadly speaking, expand as the nuclear evolution of their cores proceeds. Hence, initially being in a relatively wide binary means that a star can fill its Roche lobe later in its nuclear evolution. In turn, stars which lose their envelopes later in their nuclear evolution generally have more massive cores at the end of their lives than those which lose their envelopes early. Then CEE allows those cores, or the products of those cores, to be found in short-period binaries.

CEE is especially important for making binaries with compact components. The timescale of angular-momentum losses by gravitational-wave emission becomes dramatically shorter for closer orbits, and only binaries that are already fairly close will be significantly modified by gravitational-wave emission during a Hubble time. While Paczyński's field-defining paper was still only available as a preprint, Smarr & Blandford (1976) wrote a manuscript that discussed CEE for the formation of the newly-discovered Hulse–Taylor pulsar (Hulse & Taylor 1975).[2] Hulse and Taylor later won the Nobel prize for using this system to demonstrate that the observed angular-momentum loss follows the predictions of general relativity for gravitational waves.

Today we have observed populations of even more spectacular gravitational wave-emitting systems than the Hulse–Taylor pulsar—mergers of black holes and neutron stars—and a Nobel Prize has been awarded for the actual detection of the gravitational waves themselves. For those gravitational-wave merger systems that are produced from isolated field binary stars, leading formation channels involve CEE. In this case the uncertainty in our predictions includes not only the uncertainty from a commonly-used energy-balance argument for the outcome of CEE (see Chapter 3), but also uncertainties regarding which systems enter CEE (see Chapter 5). Moreover, even for systems for which the energy-balance argument would predict successful ejection, it is not known which stellar structures allow envelope ejection (see Chapters 8 and 9).

Another Nobel prize-associated stellar phenomenon is commonly expected to involve CEE during the evolution of the binary stars which produce it. Type Ia supernovae (SNe Ia), which were used to infer the existence of cosmological "dark energy," are thought to be produced by the explosion of carbon–oxygen white dwarfs in particular binary configurations. Astrophysicists still argue about the most likely progenitors for the diverse population of SNe Ia, but the dominant evolutionary scenarios for the production of these explosions typically involve CEE, as schematically depicted in Figure 1.2. This is broadly because wide initial binaries allow one (or both) of the stars to produce a massive carbon-oxygen core. However, to allow later accretion, or merger, it helps if the stars are, after the formation of the white dwarf(s), brought much closer than they were for those wide initial orbits. One of the most highly-cited papers from the history of the development of common-envelope theory, Webbink (1984), studied populations produced by the mergers of white-dwarf systems, partly applied to SNe Ia. Webbink (1984) is credited with formalizing an energy-balance argument for common-envelope ejection, which has been widely applied to studying populations of binary stars (for more on the modern context for energetic arguments about CEE, see Chapter 3).

However, it is not only rare stellar exotica and explosive transients for which CEE is important. Regulus (α Leonis) is a first magnitude star, known and named millennia ago, that was only recently discovered to be a binary star (Gies et al. 2008; the 0.3 $M_{\odot}$ companion is probably a low-mass white dwarf). The future evolution of the Regulus binary depends heavily on the outcome of a CE phase and illustrates common evolutionary paths for many "normal" Galactic binaries (see Figure 1.3; based on Rappaport et al. 2009). This post-mass-transfer binary system is expected to enter a common-envelope phase when the presently more-luminous star leaves the

[2] Note that Smarr and Blandford describe the mechanism as "double-core evolution." That term was common in the early days of work on the common-envelope process. However, "double-core evolution" is now generally used with a more specific meaning (see Sections 5.8 and 10.2.3). This can make reading early papers confusing.

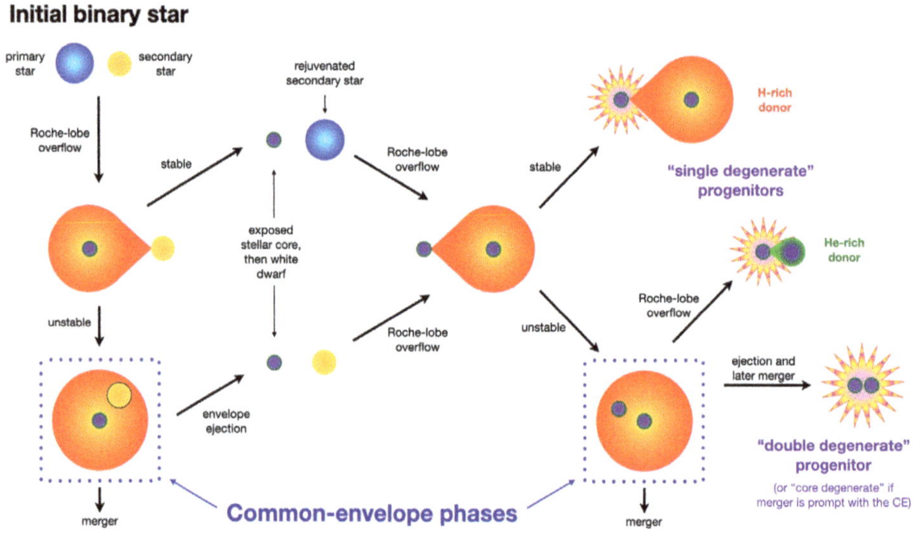

Figure 1.2. It is not certain how the universe produces the observed diversity of Type Ia supernovae—thermonuclear explosions of carbon–oxygen white dwarfs—but the most widely studied origin channels involve common-envelope evolution.

main sequence. Depending on the outcome of the common-envelope phase, this binary may evolve into quite a diverse range of configurations. If future common-envelope ejection is successful in this system, the exact post-common-envelope orbital period will qualitatively determine the future evolution, including the *direction* of the post-common-envelope mass transfer. This is because a sufficiently short post-common-envelope orbital period will enable gravitational-wave radiation to bring the binary into contact while the exposed post-common-envelope core is still burning helium, in which case that star will be the less-dense component and so will be the mass donor. If the post-common-envelope orbital separation is wide enough that gravitational-wave losses take longer than the nuclear-burning lifetime of the exposed core to bring the system into contact, then both components will have become degenerate, in which case the less-dense component will be the less-massive white dwarf (cf Rappaport et al. 2009). Both of these potential futures for Regulus would be observationally classified as AM CVn-type systems, but they would have different detailed properties. So in this case it is not sufficient to know whether the common-envelope phase leads to a successful envelope ejection; the post-common-envelope period also makes a qualitative difference to the future evolution. Note also that the potential final fates described for surviving post-common-envelope binaries in Figure 1.3 are uncertain partly because of uncertain mass-transfer stability, another topic intrinsically linked with CEE.

Another possibility for the future of Regulus and similar binaries is a common-envelope phase which leads to a merger, also illustrated in Figure 1.3. Common-envelope modelers not only wish to know whether the common-envelope phase leads to a merger or to envelope ejection, but also to predict the appearance of any such

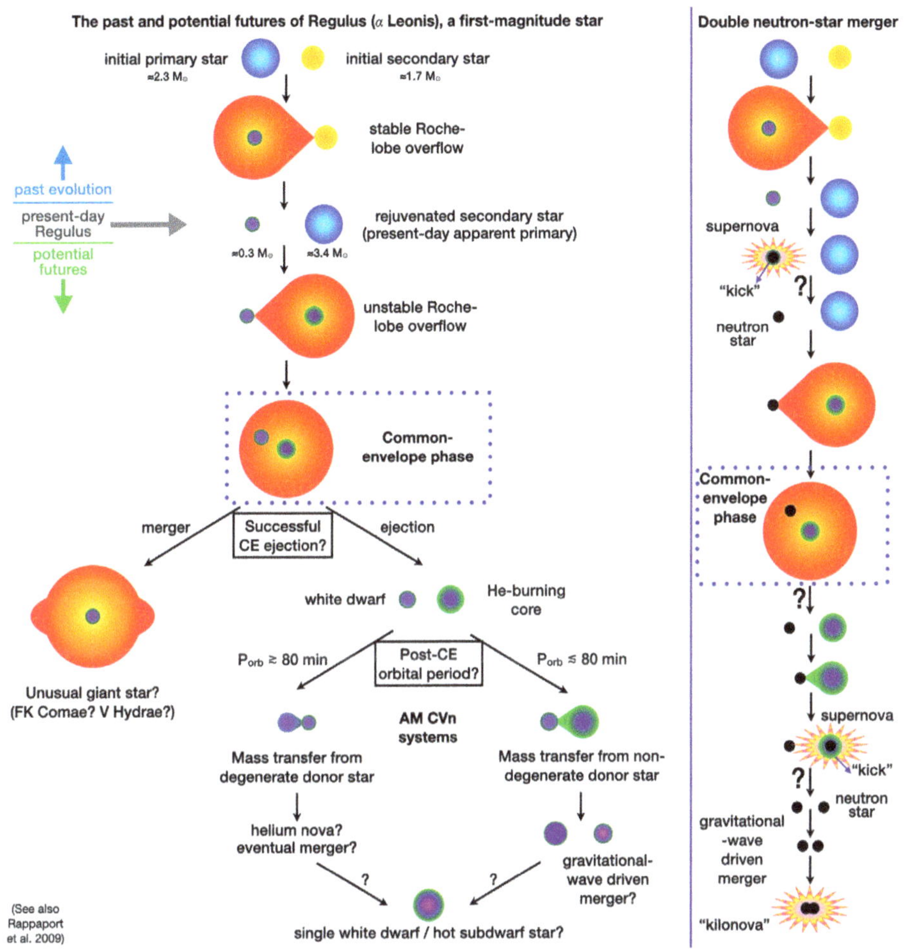

Figure 1.3. The left-hand, larger part of the schematic illustrates the uncertainty introduced in the predicted future evolution of Regulus by not being able to predict the outcome of common-envelope evolution (closely based on Rappaport et al. 2009). Taking the rightmost of the forking paths for the future of Regulus describes a future which is qualitatively similar to the standard single-common-envelope scenario for the formation of close double neutron star (DNS) binaries, as illustrated in the right-hand part of the schematic. For this route to DNS formation we do not show the branching alternatives; the supernovae introduce another significant uncertainty in the future of any individual system.

post-merger star. Rappaport et al. (2009) suggest two archetypes for rapidly-rotating giant stars, V Hydrae and the FK Comae class, which may be examples of common-envelope mergers for that possible future of Regulus and similar binaries. To model these possibilities we would want to be able to predict the angular momentum distribution of the merger product, and ideally model any changes in the composition of the giant's envelope due to the common-envelope phase and merger.

The potential evolutionary paths for Regulus may, eventually, lead to a gravitational-wave driven compact-object merger. As illustrated in Figure 1.3, this potential future is qualitatively similar to the canonical formation channel for double neutron-star merger systems (see, e.g., Dewi & Pols 2003; Ivanova et al. 2003; Tauris et al. 2017). Mergers of double neutron stars are thought to be responsible for the formation of some stellar-mass gravitational-wave merger sources, "kilonovae," "short-hard" gamma-ray bursts, and for a significant fraction of the *r*-process nucleosynthesis in the universe (see, e.g., Abbott et al. 2017a, 2017b).

The above are only some examples of the importance of CEE in our present understanding of the formation of a diverse range of stellar systems. Many observed classes of close binaries that contain one compact object, or stripped-envelope exposed core, are also thought to have formed via CEE. This includes the broad class of "cataclysmic variables"—containing one white dwarf and a low-mass main-sequence star in a sufficiently close orbit for magnetic braking or gravitational-wave radiation to drive the mass transfer. The system which inspired Paczyński's 1976 paper, V471 Tau, is expected to evolve into a cataclysmic variable.

The origin of short-period low-mass black hole X-ray binaries is more debatable—for some common-envelope parameters and physical assumptions they are formed by population models in adequate numbers to explain the observations; for other assumptions they are not. The mergers of black hole binaries that have been detected by gravitational waves, especially the more massive examples, are another debated class of system: for some assumed properties of mass-transfer instability and common-envelope physics their production may be dominated by isolated binary evolution involving CEE, but alternative possibilities exist. We shall return to comparisons between observed populations of systems and common-envelope models in Chapter 10.

Another reason that study of CEE is especially exciting at present is that the relevant observational evidence may no longer be entirely based on interpreting the prior evolution of observed post-common-envelope systems, with some observed transients potentially associated with common-envelope events (see Section 10.5).

Finally on the subject of why understanding the physics of CE is broadly important, we note that CEE may also control the fate of *single stars*—not post-merger single stars, but stars that by standard definitions are born as single stars. Many such single stars are in multiple systems with substellar objects, companions too low-mass to fuse protons into helium, i.e., planets or brown dwarfs. To give a concrete example: if, after the Sun evolves into a giant star, would it meaningfully affect the future evolution of the Sun if Jupiter were to spiral into the giant envelope? How much does the answer to that question depend on the precise stage at which the extreme-mass-ratio common-envelope episode begins, and how much would the outcome be affected if Jupiter were ten or thirty times more massive? The potential influence of CEE involving substellar objects on the evolution of single stars has been discussed by, e.g., Soker (1998), Nelemans & Tauris (1998), Soker & Harpaz (2000), and recent modeling efforts include Staff et al. (2016) and Kramer et al. (2020).

1.3 Why Is Modeling Common-envelope Evolution Difficult?

Given the importance of CEE, and the fact that it has been studied for over 40 years, readers might be asking themselves why so many fundamental questions about the physics and phenomenology related to common-envelope events remain unanswered. We hope that this book will at least help to give clarity about why our detailed knowledge of CEE remains so uncertain.

Most of this book examines how we can appropriately model the complex (astro-) physical process of CEE through both analytic arguments and computational simulations. Theoretical astrophysics is often about deciding which approximations can be made while still producing a meaningful answer. CEE is a good test case for learning this skill, despite the fact that we do not yet collectively agree about which approximations are optimal. The full problem involves diverse physics, is inherently multidimensional, and involves a wide range of timescales. For the important unsolved question of "which systems *enter* common-envelope evolution?" we are still relegated to one-dimensional approximations, which strictly speaking break down before the systems enter common-envelope evolution.

Perhaps it is not obvious to the reader why CEE is such a hard problem: "why is modeling two, interacting, stars not roughly twice as difficult as modeling one star?" Leaving aside the dubious premise that accurate models for single-star evolution are easy, perhaps the following illustration helps. Take a somewhat extreme case, one which is highly topical and by no means ridiculous, of a red supergiant entering CEE with a low-mass black hole. The red supergiant has a radius of order 1000 $R_\odot$ ($\approx 7 \times 10^{11}$ m), the black hole of order 10 km. So there are almost eight orders of magnitude in difference between the radius of that black hole and the red supergiant. This is a similar difference in size as between that red supergiant and a 2 kpc galactic bulge (i.e., $\approx 6 \times 10^{19}$ m), similar to the bulge of the Milky Way! A brute-force hydrodynamical simulation of such a system would require following the evolution in 3D of $(10^8)^3 = 10^{24}$ mass elements. Such a task is far beyond our computational capabilities and will remain so for the foreseeable future. Hence even approximate solutions to this problem require careful deployment of computational resources and thus provide an excellent example of a challenging modern computational astrophysics problem.

While the comparison in Figure 1.3 between low-mass and high-mass binary-star evolutionary scenarios is useful in illustrating parallels between different classes of system, and the central importance of CEE, we stress that common-envelope phases cannot be easily rescaled between low-mass and high-mass systems. CEE is not a scale-free problem, so what we learn from any particular high-fidelity, expensive computational model can only be extrapolated very carefully.

We do not have codes that can, in practice, evolve a common-envelope phase from beginning to end. The convention is to break CEE into different conceptual phases for which different physics is thought to be dominant. Different phases are more suitable for study using different types of code, i.e., with different approximations. Of course these phases are not discretely separated in nature, but understanding them is a good way to obtain an overview of CEE and the approximations we make when modeling common-envelope events. These phases will be the subject of the next Chapter.

References

Abbott, B. P., Abbott, R., Abbott, T. D., et al. 2017a, ApJL, 848, L13
Abbott, B. P., Abbott, R., Abbott, T. D., et al. 2017b, ApJL, 848, L12
Dewi, J. D. M., & Pols, O. R. 2003, MNRAS, 344, 629
Gies, D. R., Dieterich, S., Richardson, N. D., et al. 2008, ApJL, 682, L117
Hulse, R. A., & Taylor, J. H. 1975, ApJL, 195, L51
Ivanova, N., Belczynski, K., Kalogera, V., Rasio, F. A., & Taam, R. E. 2003, ApJ, 592, 475
Kramer, M., Schneider, F. R. N., Ohlmann, S. T., et al. 2020, A&A, 642, A97
Nelemans, G., & Tauris, T. M. 1998, A&A, 335, L85
Paczynski, B. 1976, in IAU Symp. 73, Structure and Evolution of Close Binary Systems, ed. P. Eggleton, S. Mitton, & J. Whelan (Dordrecht: Reidel), 75
Rappaport, S., Podsiadlowski, P., & Horev, I. 2009, ApJ, 698, 666
Smarr, L. L., & Blandford, R. 1976, ApJ, 207, 574
Soker, N. 1998, AJ, 116, 1308
Soker, N., & Harpaz, A. 2000, MNRAS, 317, 861
Staff, J. E., De Marco, O., Wood, P., Galaviz, P., & Passy, J.-C. 2016, MNRAS, 458, 832
Tauris, T. M., Kramer, M., Freire, P. C. C., et al. 2017, ApJ, 846, 170
Webbink, R. F. 1984, ApJ, 277, 355
Webbink, R.F. 1975, PhD thesis, Univ. Cambridge

Natalia Ivanova, Stephen Justham and Paul Ricker

Chapter 2

Main Phases

A common-envelope evolution (CEE) can be seen as a sequence of five physical phases, in which each of the phases takes place over a separate timescale and is driven by different physics: (i) loss of orbital stability and the onset of the common envelope (for brevity, we will refer to this stage as to LOS), (ii) the plunge-in; (iii) self-regulated spiral-in; (iv) delayed ejection; (v) post-CE evolution. Not all phases may occur in every common-envelope system. We show schematically the main phases in Figure 2.1.

Because the transitions between phases are gradual, it is often difficult to identify well-motivated quantitative criteria to determine when a simulation moves from one to the next. Thus much current work proceeds from qualitative estimates of transition between phases. However, in order to compare work by different groups, quantitative measures are needed, though they may be somewhat arbitrary. In the past, misclassification of phases and the important physics has been a source of confusion and misunderstanding. Here we delineate some of the considerations needed to overcome this confusion.

2.1 Characteristic Timescales

Before giving descriptions of the different phases, we clarify the definitions for timescales that we use throughout the book.

The global dynamical timescale $\tau_{\rm dyn}$ is the characteristic timescale on which the object under consideration responds to a mechanical perturbation. This timescale can change significantly during a common-envelope episode.

There is no unique definition of the dynamical timescale. Here we give some examples of how this timescale might be estimated. While the stars are still two separate objects, the dynamical timescale is commonly estimated to be of order one orbital period:

doi:10.1088/2514-3433/abb6f0ch2

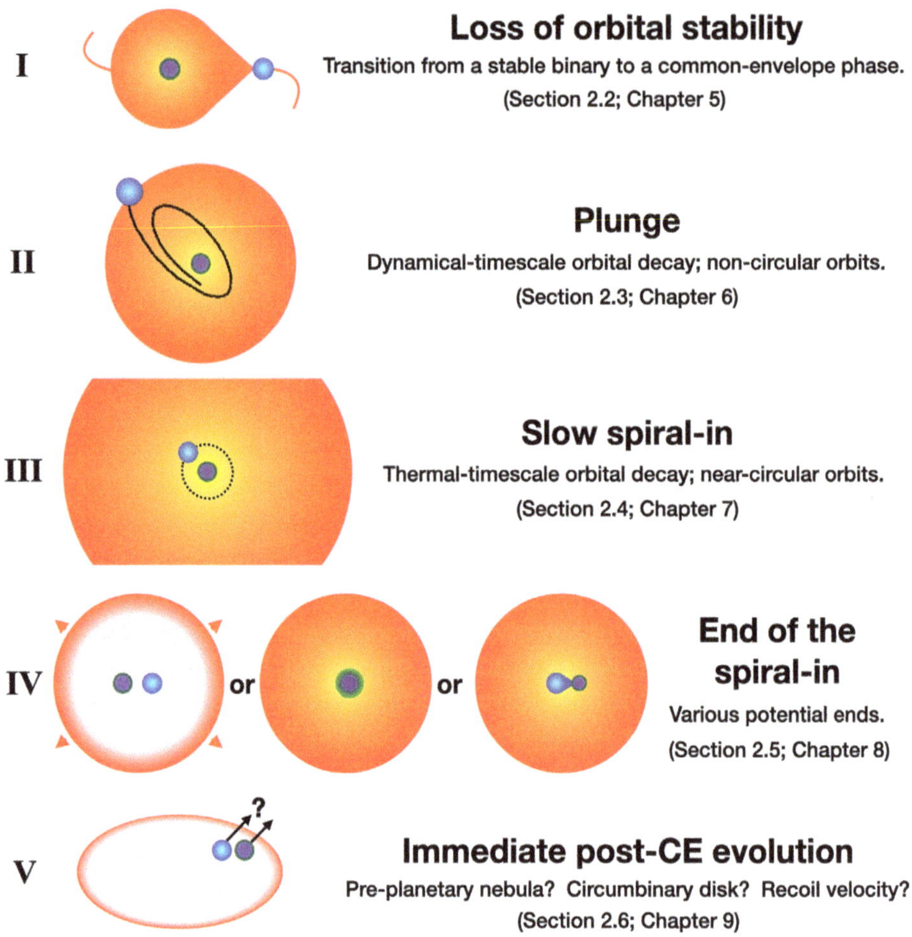

Figure 2.1. The main conceptual phases of a common envelope event. Not all phases are expected to take place in every common envelope event.

$$\begin{aligned}\tau_{\rm dyn,orb} = P_{\rm orb} &= 2\pi\sqrt{\frac{a^3}{G(M_d + M_{\rm comp})}} \\ &\approx 30\ {\rm min}\sqrt{\frac{(a/R_\odot)^3}{(M_d + M_{\rm comp})/M_\odot}}.\end{aligned} \tag{2.1}$$

Here M_d and M_2 are the masses of the donor and the companion, and a is the current orbital separation.

When the two stars are orbiting inside the common envelope, the global dynamical timescale is of order the free-fall time of the entire envelope. This is commonly estimated by comparison with the collapse timescale of a uniform-density gas sphere:

$$\tau_{\rm dyn,CE} = \frac{\pi}{2^{3/2}}\sqrt{\frac{R_{\rm CE}^3}{GM}} \approx 30 \text{ min } \sqrt{\frac{(R_{\rm CE}/R_\odot)^3}{M_{\rm tot}/M_\odot}}. \tag{2.2}$$

Here $M_{\rm tot}$ is the mass that is currently located within the common envelope, inclusive of both the stars but not the material ejected from the donor star's envelope during any of the previous phases. We distinguish $M_{\rm tot}$ from the mass of the common envelope itself, which includes bound and unbound gas. $R_{\rm CE}$ is the radius of the common envelope at a given time. For a subgiant with mass of 1 $M_\odot$ and radius of 3 $R_\odot$ and a companion of a smaller mass, this estimate gives $\tau_{\rm dyn,CE}$ of only a few hours, while for an asymptotic giant branch (AGB) donor of a similar mass and a radius of 200 $R_\odot$, the estimate gives about 60 days.

A common rough expression for $\tau_{\rm dyn}$ is $\sqrt{1/G\bar{\rho}}$, where $\bar{\rho}$ denotes the mean density. When the mean density decreases during the CEE, $\tau_{\rm dyn,CE}$ typically increases by orders of magnitude over its initial value. However the orbital timescale $\tau_{\rm dyn,orb}$ decreases, as depicted in Figure 2.2.

Another way to approach the global dynamical timescale is by considering an adiabatic perturbation and the time that it takes for the resulting sound wave to travel from the center to the surface of the star. This defines the *global sonic timescale*, $\tau_{\rm dyn,son}$. For a non-degenerate star in which pressure is mainly due to an ideal gas, $\tau_{\rm dyn,son}$ is of the same order as the global dynamical time as defined through the free-fall time (see, e.g., Section 1.4 in Hansen et al. 2004).

The *local dynamical timescale* is the time it takes for an adiabatic sound wave to travel from some radius r_l to the surface of the common envelope:

$$\tau_{\rm dyn,loc}(r_l) = \int_{r_l}^{R_{\rm CE}} \frac{dr}{c_s(r)}. \tag{2.3}$$

Here $c_s(r)$ is the sonic velocity at a distance r from the center of the envelope. With the definitions adopted here, $\tau_{\rm dyn,son} = \tau_{\rm dyn,loc}(0)$.

The *thermal timescale* $\tau_{\rm th}$ is the time during which the envelope can radiate away all its current internal energy at its present surface luminosity,

$$\tau_{\rm th,CE} = \frac{1}{L_{\rm CE}} \int_{CE} u(m)dm. \tag{2.4}$$

In the same way as for the sonic timescale, this timescale can be defined for the region external to a given mass coordinate as well as globally. In the above equation, L_{CE} is the current surface luminosity of the common envelope. This luminosity, as will be discussed later, can vastly exceed the initial value of the donor star's luminosity. Initially, this timescale is similar to the star's Kelvin–Helmholtz timescale, $\tau_{\rm KH} \approx 2 \times 10^7 \text{ yr} \times (M/M_\odot)^2(R/R_\odot)^{-1}(L/L_\odot)^{-1}$. (Often the thermal timescale of the envelope is more relevant than the whole-star timescale, for which replace $(M/M_\odot)^2$ with $(M/M_\odot)(M_{\rm env}/M_\odot)$ in that expression, with $M_{\rm env}$ the mass of the envelope.) The initial thermal timescale varies; it can be as long as about 50,000 years for 1 $M_\odot$ giant of 10 $R_\odot$, and as short as less than a year for an evolved AGB star of 10 $M_\odot$. In the latter, the thermal timescale is already comparable to the dynamical timescale in the unperturbed star.

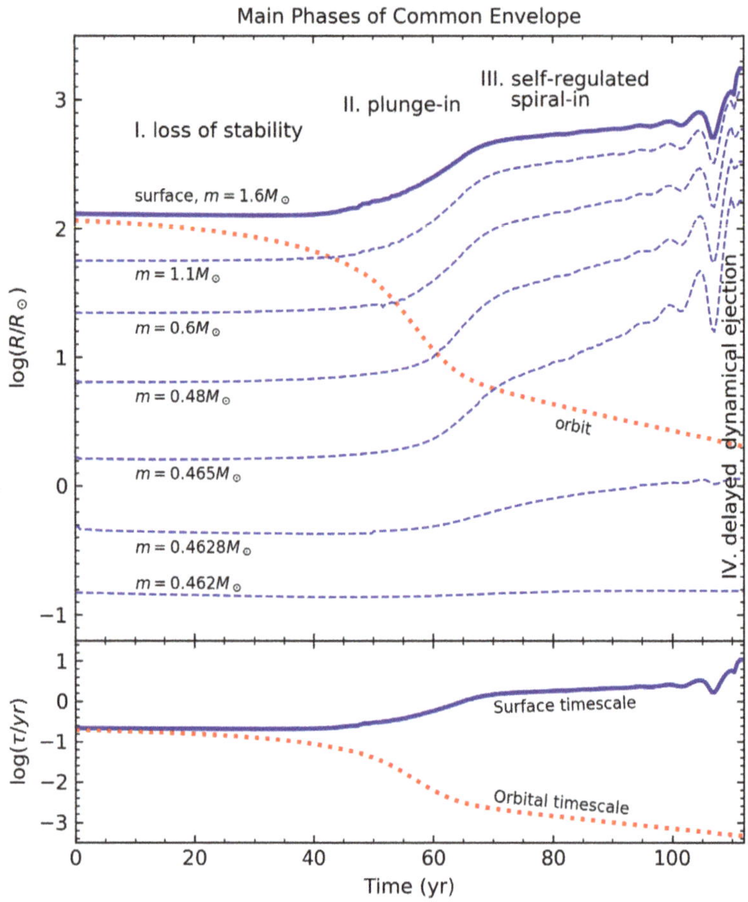

Figure 2.2. The main phases of a common envelope event, together with the evolution of relevant dynamical timescales. This example is for a 1.6 $M_\odot$ red giant and a 0.3 $M_\odot$ white dwarf, using data from one-dimensional hydrodynamical simulations in Ivanova (2002). Not all of these phases are expected to happen during all common envelope events; timescales can vary widely. The blue solid line indicates the surface, which here means the outer boundary of the bound envelope. The dashed blue lines represent locations at fixed mass coordinates of the envelope (not taking into account the mass of the secondary). The dotted line shows the location of the inspiraling secondary.

Both the radius and luminosity grow during CEE, so $\tau_{\rm th,CE}$ decreases by orders of magnitude in comparison with its initial value. As with an evolved AGB star, an extended common envelope may have a new thermal timescale that is shorter than its new dynamical timescale, $\tau_{\rm th,CE} \leqslant \tau_{\rm dyn,CE}$. Under such conditions one must consider both dynamical and thermal evolution in predicting the CEE outcome.

The *thermal adjustment timescale* $\tau_{\rm adj}$ is the time required to redistribute the heat from a new extra energy source throughout the envelope, bringing the star into a new thermal equilibrium with that heat source.[1] While this timescale is often confused with the thermal timescale, it characterizes how fast energy may be transported after a temperature perturbation rather than how much energy is flowing through the unperturbed configuration. In the case of a common envelope, this timescale would describe the time during which the envelope would change (expand) from its initial state if the orbital energy or other energy were released inside of it. The resulting timescale will depend on the convective velocity. For example, in low-mass giants, the convective velocity v_c is a few per cent of the local sonic velocity c_s, so $\tau_{\rm adj}$ should be longer than the global dynamical timescale by at least a factor of about c_s/v_c.

The *collisional timescale* $\tau_{\rm coll}$ is the mean time between two successive scatterings of a given particle. This timescale is related to the timescale for reaching local thermodynamic equilibrium (LTE), such that that the gas can be described by a Maxwell–Boltzmann distribution of temperature T. The time to come to local equilibrium is naturally larger than $\tau_{\rm coll}$, as it should take at least a few collisions to come to equilibrium. $\tau_{\rm coll}$ can be estimated as

$$\tau_{\rm coll} \approx \frac{\lambda_f}{v_p} \approx \frac{1}{\sqrt{2}\,\sigma n}\frac{1}{v_p}. \tag{2.5}$$

Here λ_f is Maxwell's mean free path, n is the particle number density, v_p is the most probable particle speed, and σ is the collision cross section. For a hydrogen ion, the cross section based on its kinetic diameter is $\sigma \approx 9 \times 10^{-15}$ cm^2. In cases for which each particle of a gas has three degrees of freedom, its sound velocity (in LTE) can be related to its most probable velocity via $c_s = \sqrt{5/6}\,v_p$; hence for a gas with mean molecular weight[2] μ and density ρ:

$$\tau_{\rm coll} = \sqrt{\frac{3}{5}}\frac{1}{\sigma n c_s} \approx 4 \times 10^{-10}\ {\rm s}\ \frac{\mu}{\rho c_s}. \tag{2.6}$$

For densities and sonic velocities typical of stellar envelopes, the collisional timescale is hence very short. Comparing Equations (2.3) and (2.6), the collisional time can become comparable to the local dynamical time when the density of the envelope becomes less than $\sim 4 \times 10^{-10}/\ell$, where ℓ is the characteristic size of the envelope at any moment during the envelope's expansion. For any envelope of about a solar mass or more, this condition cannot be achieved until ℓ becomes larger than $\sim 10^{21}$ cm.

---

[1] A star is not in thermal equilibrium with its environment! Nonetheless, this terminology of a star being in global "thermal equilibrium" is common. This description indicates that the thermal structure of the star is not changing on a thermal timescale, and only changing on the evolutionary timescale, i.e., typically the timescale of nuclear evolution. More formally we might write that the specific entropy, s, is nearly constant in time at all points through the star (i.e., $\int |ds/dt| dm \approx 0$). Sometimes this state is called "gravothermal equilibrium," since the local heat content of the stellar matter is (approximately) not changing, and the structure is (approximately) neither expanding nor contracting.

[2] "Mean molecular weight" is more terrible legacy terminology. Physicists are trained to clearly distinguish mass from weight, but μ does not represent a weight; it describes a mass. Moreover, in the context of stars μ only extremely rarely refers to molecules. This is a mean particle mass.

2.2 Phase I: The Loss of Orbital Stability and the Onset of the Common Envelope

During this phase the progenitor binary is transformed from a stable, likely non-eccentric binary, where the donor also likely was synchronized with the orbit, into a spiraling-in binary with a common envelope surrounding it. This stage precedes what is most often meant by the term "common envelope," and it is not often considered by 3D hydrodynamical simulations. During the loss of stability (LOS) stage, the two stars can be treated as physically distinct objects. The companion orbits either outside of the future CE, or inside the donor's expanded, rarefied outer layers. The orbit of the companion may enter inside the common envelope either because the donor expands or because the orbit shrinks, or a combination of both. The expansion of the donor would typically be rapid (faster than the global thermal timescale, while slower than the global dynamical timescale).

Engulfment by shrinking of the companion's orbit can only take place if the orbital angular momentum $J_{\rm orb}$ is decreasing rapidly, either due to loss of angular momentum from the system with the mass outflow through the outer Lagrangian points Ivanova & Nandez (2016), or the initial $J_{\rm orb}$ is transferred into the donor's rotation in the course of development of Darwin instability (Darwin 1879; Hut 1980; Lai et al. 1993), or a secular tidal instability in triple systems (Eggleton & Kiseleva-Eggleton 2001).

The LOS phase is typically considered to begin only after the donor overfills its Roche lobe (here and later it is denoted as RLOF, which stands for Roche-lobe overflow); accordingly the mass transfer starts during the LOS phase. The beginning of the phase may involve a thermal-timescale mass transfer that could last hundreds of years (e.g., Podsiadlowski et al. 2002). In that case, mass transfer is likely to strongly affect the donor's structure. Either during the thermal timescale mass transfer, or during the mass outflows, a substantial fraction of the initial donor's envelope mass can be lost.

At the start of the LOS phase, a binary would have orbital separation a_i and the donor would have mass $M_{d,i}$. The donor's radius at the start of this phase, $R_{d,i}$, can be related to the orbital separation through one of the approximations for the effective Roche-lobe radius $r_{\rm d,l}$, e.g., Eggleton (1983)

$$R_{d,i} = r_{\rm d,l} a_i = \frac{0.49 q^{2/3}}{0.6 q^{2/3} + \ln(1 + q^{1/3})} a_i, \quad q = M_{d,i}/M_{\rm comp}. \tag{2.7}$$

However, at the end of the LOS stage, neither the orbital separation nor the total mass of the binary is expected to be the same as they were at the start of the LOS stage. The Roche approximation is also not expected to be valid; it is valid only when the material orbits synchronously with the binary rotation, and that assumption breaks down.

During the LOS phase, the change in the orbital separation a can be assumed to be less than one per cent when averaged over an orbital period $P_{\rm orb}$. When $|(\dot{a} P_{\rm orb})|/a \approx 0.1$, the system has clearly entered the dynamical plunge-in phase, as described in the next section.

To summarize, a heuristic way to describe the LOS phase is as follows:

- *Duration:* thermal timescale to several dynamical timescales.
- *Driving mechanisms:* mass and orbital angular momentum loss.
- *Quantities transformed:* donor mass, binary separation, angular momentum.
- *End:* the orbital decay becomes dynamical, $|(\dot{a}P_{\rm orb})|/a \approx 0.1$.

For more technical details see Chapter 5.

2.3 Phase II: The Plunge-in

Often referred to as a spiral-in, this is the fastest phase of orbital shrinkage. The companion orbits within the donor's envelope, depositing some of its orbital energy $E_{\rm orb}$ in the envelope and driving expansion of the envelope. The plunge-in takes place over several orbital periods, and can be faster than one initial period.

The plunge-in may end with ejection of the entire envelope, with merger of both stars, or transition to a slower phase of spiral-in (see Section 2.4 and Chapter 7). A part of the envelope is always ejected, even in the case of a merger.

If an envelope ejection happens during the plunge-in, it would take place on a dynamical timescale (see Section 8.2 for more details). This rapid process is what has often been thought of as "common-envelope ejection," but it is not the only possible way to remove the envelope.

During the plunge-in phase, the orbit is not Keplerian, since the mass inside the companion's trajectory changes even during one revolution of the companion around the donor's core. The rate of decay of the orbital energy can provide a useful criterion to indicate the transition into a slow spiral-in:

$$(\dot{E}_{\rm orb}P_{\rm orb})/E_{\rm orb}| \leqslant 0.01. \tag{2.8}$$

(Because the orbit is not closed, here $P_{\rm orb}$ refers to the radial period.) The plunge-in stage is often treated as purely dynamical, i.e., energy losses from radiation, or energy transport by convection, are neglected. This is a reasonable approximation as long as the convection is subsonic and the thermal timescale of the envelope is much longer than the dynamical timescale. This approximation may be problematic for AGB and supergiant stars, which often have similar thermal and dynamical timescales. It is not clear that these stars would have a well-defined plunge-in phase.

Most numerical three-dimensional (3D) simulations of common envelope events so far have been restricted to modeling the plunge-in stage, neglecting radiative energy losses.

In brief, we can characterize the plunge-in phase as follows:

- *Duration:* several *current* dynamical timescales; the entire spiral-in can proceed within less than the initial orbital period.
- *Driving mechanisms:* mechanical energy transfer from the binary orbit into the common envelope.
- *Quantities transformed:* the separation between the companion and the core of the donor and the mass of the envelope; at least a part of the envelope is always ejected, even for mergers.

- *End:* a merger, or transition to a slower spiral-in when the energy transfer from the orbit becomes small:

$$(\dot{E}_{\rm orb}P_{\rm orb})/E_{\rm orb}| \leqslant 0.01. \tag{2.9}$$

Clean envelope ejection during the dynamical plunge-in, leading to a surviving post-CE binary, is expected to take place in at least some systems, but so far this has not been observed in simulations.

For more technical details see Chapter 6.

2.4 Phase III: The Slow Spiral-in

The envelope may expand enough so that the binary formed by the core and companion finds itself in a sufficiently expanded envelope that the rate of orbital decay due to global tides decreases. The dominant orbital decay mechanisms change to local effects, including dynamical friction (Section 7.2.4). For this reason it has been commonly assumed that all orbital energy release during this stage can be described by a so-called “frictional” luminosity $L_{\rm fr}$, which acts to heat the envelope material where it is deposited, increasing the envelope’s entropy. The assumption was first made in Meyer & Meyer-Hofmeister (1979), and in Chapter 7 we will discuss recent criticism of this assumption.

The expanded common envelope has a higher surface luminosity $L_{\rm CE}$ than the initial donor. In some cases, the frictional luminosity $L_{\rm fr}$ becomes low enough for the common envelope to transport all the energy to the surface, where the energy is radiated away in steady state. On the other hand, the rate of spiral-in might be instantaneously determined by the local density in the vicinity of the orbit. In this case, too much (too little) instantaneous heating causes the local density to decrease (increase), decreasing (increasing) the rate of spiral-in and therefore of heating. The balance of the two effects may lead to the formation of a transitory *self-regulating* state (Meyer & Meyer-Hofmeister 1979), during which all released orbital energy is transported to the surface and is radiated away. In Chapter 7 we will discuss criticism of this assumption. Whether the ideally balanced self-regulated regime is achieved or not, the slow spiral-in phase is not secularly stable; its termination has a different physics from the slow spiral-in itself and will be given special attention separately (see Section 2.5).

This phase operates on the *thermal timescale* of the expanded common envelope. This affects the numerical methods which are suitable for meaningfully simulating the evolution during the slow spiral-in. The 3D codes which have so far been used to simulate CEE do not treat the energy losses due to radiation, and all current 3D hydro codes have trouble in performing calculations of an adequate duration (see Chapter 4). However, one-dimensional (1D) methods are not self-consistent in where and how the energy is deposited (see Section 7.4).

We summarize the slow spiral-in as follows:

- *Duration:* new thermal timescale of the expanded common envelope. Note that this time can be comparable to the envelope’s new dynamical timescale.

Important: transitional binary orbital period is significantly smaller than the dynamical timescale of the expended common envelope.

- *Driving mechanisms:* "frictional" energy deposition, transport through the envelope of the energy liberated from the orbital to the surface, energy loss from the surface of the envelope.
- *Quantities transformed:* the entropy of the expanded envelope; some of the mass of the envelope can be lost through one or more "shell"-triggered ejections (see Section 8.4).
- *End:* instability and ejection of the envelope; merger; Roche-lobe overflow of the companion onto the core of the donor; instability leading to a second plunge-in or strong non-radial pulsations. (For more, see Section 2.5)

For more technical details see Chapter 7.

2.5 Phase IV: Termination of the Slow Spiral-in Phase

The slow spiral-in state is not stable forever, and the end is expected to take place on the dynamical timescale of the expanded envelope. One possible end is the ejection of the envelope, e.g., via delayed dynamical ejection (Ivanova 2002; Han et al. 2002). Alternatively, the common-envelope event can terminate with a merger, when either of the secondary or core of the primary overfills its Roche lobe. The merger may last hundreds or more orbits of the stellar cores, and is termed in that case a "slow" merger (Ivanova 2002; Ivanova & Podsiadlowski 2003). The merger may also provide an alternative energy source to eject the envelope around the merged secondary and the core of the primary, due to explosive nuclear burning during the merger, e.g., when two CE white dwarfs merge, or a main sequence companion injects its hydrogen into the helium-burning shell layer of the more evolved donor (Ivanova et al. 2002; Ivanova & Podsiadlowski 2003; Podsiadlowski et al. 2010).

A slow spiral-in ("phase III") could, in principle, also be followed by another dynamical plunge ("phase II") if the mechanism maintaining self-regulation somehow ends. For example, if the not-yet unbound envelope follows a parabolic trajectory and re-collapses on its dynamical timescale, this could increase the density around the orbit and so increase the rate of orbital decay (Ivanova & Nandez 2016). That new plunge-in phase could in turn be followed by further envelope ejection and another self-regulated phase. In principle, we know no first-principles reason why the sequence of phases could not continue as I–II–III–II–III–[···]–IV in some cases, accompanied by partial envelope ejections during each cycle (Clayton et al. 2017).

In brief, the termination of the slow spiral-in phase is characterized by:

- *Duration:* several dynamical timescales of the expanded envelope.
- *Driving mechanisms:* dynamical instability of the expanded envelope, or the final merger of the stellar cores with a nuclear runaway in some cases.
- *End:* a clean binary, or a merged product which may be surrounded by all or part of the envelope.

2.6 Phase V: Post-CE Evolution

It is quite tempting to compare the final properties of simulated post-CE binary systems which successfully eject their envelopes at the end of phase II, or phase IV, with observed systems that are presumed to be post-common-envelope. However, even complete unbinding of the common envelope does not guarantee that the transformed binary will not change any more. For example, thermal evolution of the remnant cores might drive further mass transfer (Ivanova 2011). Winds from the remnant cores will widen the system. Any remaining circumstellar matter, which might include a circumbinary disk, might further change the orbital separation of a post-CE binary (Kashi & Soker 2011). The expanding envelope may continue its evolution into the planetary nebula phase, during which the final shape will depend on the specifics of the ejection as well as on plausible magnetic fields (Tocknell et al. 2014; Ohlmann et al. 2016; García-Segura et al. 2018; Frank et al. 2018; Ivanova & Nandez 2018; Zou et al. 2020; Reichardt et al. 2019; García-Segura et al. 2020).

The plethora of the possible post-CE events, each of which can interact with the newly-formed binary on a different timescale, from that of the orbit to that of the nuclear evolution, cannot be treated by a single numerical method. One of the most interesting and recently high-profile ways to constrain modeled common-envelope events is the comparison of simulated post-CE remnants with observed systems. However, this rich field of interesting consequences is mainly beyond the scope of this book.

References

Clayton, M., Podsiadlowski, P., Ivanova, N., & Justham, S. 2017, MNRAS, 470, 1788
Darwin, G. H. 1879, RSPS, 29, 168
Eggleton, P. P. 1983, ApJ, 268, 368
Eggleton, P. P., & Kiseleva-Eggleton, L. 2001, ApJ, 562, 1012
Frank, A., Chen, Z., Reichardt, T., et al. 2018, Galax, 6, 113
García-Segura, G., Ricker, P. M., & Taam, R. E. 2018, ApJ, 860, 19
García-Segura, G., Taam, R. E., & Ricker, P. M. 2020, ApJ, 893, 150
Han, Z., Podsiadlowski, P., Maxted, P. F. L., Marsh, T. R., & Ivanova, N. 2002, MNRAS, 336, 449
Hansen, C. J., Kawaler, S. D., & Trimble, V. 2004, Stellar Interiors: Physical Principles, Structure, and Evolution (Berlin: Springer)
Hut, P. 1980, A&A, 92, 167
Ivanova, N. 2002, DPhil thesis, Balliol College, Oxford
Ivanova, N. 2011, ApJ, 730, 76
Ivanova, N., & Nandez, J. 2018, Galax, 6, 75
Ivanova, N., & Nandez, J. L. A. 2016, MNRAS, 462, 362
Ivanova, N., & Podsiadlowski, P. 2003, in From Twilight to Highlight: The Physics of Supernovae, ed. W. Hillebrandt, & B. Leibundgut (Berlin: Springer), 19
Ivanova, N., Podsiadlowski, P., & Spruit, H. 2002, MNRAS, 334, 819
Kashi, A., & Soker, N. 2011, MNRAS, 417, 1466
Lai, D., Rasio, F. A., & Shapiro, S. L. 1993, ApJL, 406, L63
Meyer, F., & Meyer-Hofmeister, E. 1979, A&A, 78, 167

Ohlmann, S. T., Röpke, F. K., Pakmor, R., Springel, V., & Müller, E. 2016, MNRAS, 462, L121
Podsiadlowski, P., Ivanova, N., Justham, S., & Rappaport, S. 2010, MNRAS, 406, 840
Podsiadlowski, P., Rappaport, S., & Pfahl, E. D. 2002, ApJ, 565, 1107
Reichardt, T. A., De Marco, O., Iaconi, R., Tout, C. A., & Price, D. J. 2019, MNRAS, 484, 631
Tocknell, J., De Marco, O., & Wardle, M. 2014, MNRAS, 439, 2014
Zou, Y., Frank, A., Chen, Z., et al. 2020, MNRAS, 497, 2855

Chapter 3

The Energy Budget

The energy budget takes the central place in the discussion of common-envelope (CE) events. The "energy formalism," based on considering the energy budget, is a shortcut that takes impatient scientists from a state they think they know well (pre-CE binary) to a state they think they know well (post-CE binary) while hiding a multitude of messy details in between. However, the devil is in the details. In this Chapter, we will dissect what the energy formalism means, the shortcomings of the implied definitions, and what physics is not accounted for. In particular, we stress that this expression of energy conservation assumes that the timescale of the common-envelope event is short compared to the thermal timescale of the system; this is not expected to be generally correct. We will end with the ledger of the energies of the states before and after the common-envelope event.

3.1 The Energy Formalism

The best known way to try to predict the fate of a common-envelope event is the **energy formalism** (van den Heuvel 1976; Tutukov & Yungelson 1979; Iben & Tutukov 1984; Webbink 1984; Livio & Soker 1988). The energy formalism is based on the fundamental physics concept that *total energy must be conserved.* However, in the energy formalism the total energy is regarded as consisting of only two parts: the orbital energy, $E_{\rm orb}$, and the binding energy of the envelope, $E_{\rm env}$. We equate the initial and final total energies, taking the final envelope binding energy to be zero (for reasons to be discussed later). Then energy conservation requires the initial (i) and final (f) total energies to be equal:

$$E_{\rm orb,i} + E_{\rm env,i} = E_{\rm orb,f} \tag{3.1}$$

This simplified approach leads to the most-often used form of the energy formalism, which is written by rearranging Equation (3.1) into the expected "energy demand," or so-called envelope binding energy, on one side of the equation, and the "available energy," or the released orbital energy, on the other side of the equation:

$$E_{\mathrm{bind}} = E_{\mathrm{orb,i}} - E_{\mathrm{orb,f}} = -\frac{GM_{\mathrm{d,i}}M_{\mathrm{comp}}}{2a_{\mathrm{i}}} + \frac{GM_{\mathrm{d,f}}M_{\mathrm{comp}}}{2a_{\mathrm{f}}}. \tag{3.2}$$

Here a_{i} and a_{f} are the initial and final binary separations, $M_{\mathrm{d,i}}$ and M_{comp} are the initial masses of the donor and the companion star, and $M_{\mathrm{d,f}}$ is the final mass of the star that lost its envelope mass $M_{\mathrm{d,env}}$. The binding energy E_{bind} here implies the minimum amount of mechanical energy required to be provided to the envelope to disperse it to infinity.

The initial energy of the envelope at the beginning of a CEE is often referred as the binding energy of the envelope. There is, however, a difference between the initial energy of the envelope used in Equation (3.1) and the binding energy E_{bind} used in Equation (3.2). There is sign confusion in the literature, but E_{bind} is formally defined as a positive quantity, i.e., the minimum energy which must be provided to the envelope to unbind it to infinity. The initial energy of the envelope is the sum of its gravitational potential, kinetic (inclusive of rotational energy), and internal energies; it is almost always a negative quantity, and not all of its components are bulk mechanical energy. (In Section 3.2.1 we will address the differences between the internal and the thermal energies.)

A common simplification is made by adopting $E_{\mathrm{bind}} = -E_{\mathrm{env,i}}$. It is a subject of discussion whether this equivalency is justified in all cases (see, e.g., Ivanova & Chaichenets 2011). The main problem with this simplification is that it suggests that raising the envelope's energy to zero will unbind it. However, in principle, an unperturbed star with $E_{\mathrm{env,i}} \geqslant 0$ can be *kinetically* stable, and the whole envelope does not need to outflow to infinity (Bisnovatyi-Kogan & Zeldovich 1967). Thus, providing the envelope with $-E_{\mathrm{env,i}}$ is a necessary but not sufficient condition to remove the envelope, and hence one should expect $E_{\mathrm{bind}} \geqslant -E_{\mathrm{env,i}}$.

To describe the energy of the envelope for a wide range of realistic stellar donors, a λ parameter was introduced (de Kool 1990):

$$E_{\mathrm{env,i}} = -\frac{GM_{\mathrm{d,i}}M_{\mathrm{d,env}}}{\lambda R_{\mathrm{d}}}. \tag{3.3}$$

Here R is the radius of the donor star at the start of the common envelope event. Controversies regarding what exactly should be included when calculating the binding energy of the envelope, and how to compute them, will be discussed in Section 3.2.

It cannot be expected, however, that all the available energy can be perfectly utilized to drive the envelope ejection. Perfectly efficient energy conversion implies that:

- The orbital energy has been perfectly transferred to the envelope, and only into the mechanical energy moving the envelope outwards. For instance, there are no shocks, the center-of-mass frame of the envelope does not change, and no rotational kinetic energy is added. (This is inconsistent with strict angular-momentum conservation.)
- The initial thermal energy of the envelope has been fully used to *relocate* the common envelope to infinity, not just to expand it.

- The spiral-in process provides the envelope with enough energy that the ejected envelope has no velocity at infinity: each parcel of matter is accelerated to its local escape velocity throughout the entire spiral-in process. This requires perfect fine-tuning.
- There are no other energy losses (e.g., due to radiation).

To account for the fact that realistic common envelope systems could not evolve in this idealized way, the concept of a *common-envelope efficiency* $\alpha_{\rm CE}$ was introduced (Livio & Soker 1988). Physically, this parameter represents what fraction of the released orbital energy is used to eject the envelope to infinity (with no remaining velocity). By its deep-into-the-roots meaning, this fraction of energy must be balanced against the binding energy of the envelope, not the initial energy of the envelope $E_{{\rm env},i}$, unless the equivalency of the two energies is justified:

$$\alpha_{\rm CE}(E_{\rm orb,i} - E_{\rm orb,f}) = E_{\rm bind}. \tag{3.4}$$

If there are no additional energy sources, the only theoretical constraint is that $\alpha_{\rm CE} \leqslant 1$. It is common to see literature which adopts "efficiencies" greater than one, implicitly assuming some extra energy input—typically unspecified.

The three Equations (3.2)–(3.4) together yield the most commonly used equation to predict the outcomes of common-envelope events—which are the final orbital separations of post-common envelope binaries $a_{\rm f}$—in binary population synthesis studies:

$$\alpha_{\rm CE}\left(-\frac{GM_{\rm d,i}M_{\rm comp}}{2a_{\rm i}} + \frac{GM_{\rm d,f}M_{\rm comp}}{2a_{\rm f}}\right) = \frac{GM_{\rm d,i}M_{\rm d,env}}{\lambda R}. \tag{3.5}$$

Often, the combination $\alpha_{\rm CE}\lambda$ is used as a single parameter when modeling a population of binaries. While $\alpha_{\rm CE}$ is a very widely-used parameter when trying to predict CEE outcomes, it is the common understanding among these who model CEEs that there can be no single $\alpha_{\rm CE}$ appropriate for all possible CEEs. New approaches will be described below.

3.2 The Energy of the Envelope

The energy of the envelope is given by

$$E_{\rm env,i} = \int_{M_{\rm d}-M_{\rm d,env}}^{M_{\rm d}} \left(-\frac{Gm}{r} + u_{\rm int}(m) + \frac{v(m)^2}{2}\right) dm = \Omega_{\rm env} + U_{\rm env} + K_{\rm env}. \tag{3.6}$$

Here $u_{\rm int}$ is the specific internal energy, and $v^2/2$ is the specific kinetic energy. $\Omega_{\rm env}$ stands for the integrated gravitational potential energy of the envelope, $U_{\rm env}$ stands for the integrated internal energy of the envelope (see more details in Section 3.2.1) and $K_{\rm env}$ stands for the integrated kinetic energy of the envelope. $M_{\rm d} - M_{\rm d,env}$ is the mass of the donor star without the envelope, which may but does not have to be equal to the mass of the donor's core prior the common envelope event, $M_{\rm d,core} \leqslant M_{\rm d} - M_{\rm d,env}$. Uncertainties in the remnant mass will be discussed in Section 9.1.

If Equation (3.6) were integrated over the entire star, it would provide the total energy of the star. However, as it is integrated only over a fraction of the star, it is not the true total initial energy of the envelope, because the total potential energy is not well-defined for a portion of the star, even though this term is used frequently. Nevertheless, it is a useful quantity if the core does not change in response to the removal of the envelope.

The term K_{env} in Equation (3.6) includes only local motion that is non-zero relative to the orbital motion; the energy of the orbital motion is already taken into account by the orbital energy term.

3.2.1 Internal Energy and Thermal Energy

The contribution of internal energy U_{env} in Equation (3.6) was first explicitly considered by Han et al. (1994); prior to this only gravitational potential energy was included when calculating the envelope's initial energy. Typically, U_{env} is of the same order of magnitude as $|\Omega_{env}|$ (by the virial theorem). Hence this positive energy, if it can be used in driving the envelope, can make a large difference to the outcomes of the envelope ejection. It is not yet proven what fraction of the thermal energy can be useful.

It is useful to understand what is included within the term "internal energy."

Let i be an index labeling a chemical species, which can be in different ionization states from neutral ($j = 0$) to fully ionized ($j = i$). We define y_{ij} to be the fraction of the ith species in the jth ionization state, with $\sum y_{ij} = 1$. Let $\varepsilon_{i,j}$ be the energy of the ith species in the jth ionization state; it is defined such that the energy level of a fully ionized ion is $\varepsilon_{i,j} = 0$. The difference in energies of the ground states of the $j - 1$ and j ionization stages for the species i is the ionization potential χ_{ij}. Thus, if the ionization potential of hydrogen is χ_H, then $\varepsilon_{1,0} = -\chi_H$. Similarly, transitions to molecules can be included. (If molecules are present then, strictly speaking, the internal energy should also include rotational and vibrational degrees of freedom.) The specific internal energy for a mixture of atoms, ions, electrons, and radiation, is then:

$$u_{\rm int} = \frac{3}{2}\frac{k_B T}{\mu m_u} + \frac{aT^4}{\rho} - \sum_i \sum_{j=0}^{i} \frac{x_i}{m_i} y_{ij} \varepsilon_{ij}. \tag{3.7}$$

Here k_B is the Boltzmann constant, a is the radiation constant, T is the temperature, x_i is the mass fraction of species i, m_i is the mass of species i, μ is the mean molecular weight per particle, and m_u is the atomic unit mass. For simplicity, Equation (3.7) does not include corrections to the electron energies due to Coulomb interactions and degeneracy, which are not likely to be significant in stellar envelopes. The ionization fractions y_{ij} are functions of density and temperature. For low densities, the ionization fraction can be found by solving the system of Saha equations for the mixture of species; at high densities the overlap of electron orbitals changes the effective ionization potential so as to make ionization easier (so-called "pressure ionization").

It is important to note that the choice of zero-energy state does not matter, as all that matters is the change of the specific internal energy of the envelope's material between the states in which it is bound to the donor and in which it has been ejected. (Often the internal energy in the ejected material is not considered; we will discuss this in Section 3.6.) It is conventional in stellar structure codes to use the fully ionized plasma as the zero energy level. Then the third term in Equation (3.7) does not have to be taken into account for most of the mass of stellar envelopes, so historically it has not been included in envelope internal energies. The first two terms are the thermal terms familiar from kinetic theory, and we will collectively label them as thermal energy $U_{\rm th}$. The third term is *potential* internal energy—the energy that can be released, but is not readily available (to balance gravity in an unperturbed star, for instance).

We end this section by noting that the original energy formalism implicitly assumes that the thermal energy of the envelope is fully utilized, while the orbital mechanical energy is not (this is what is implied by the use of $\alpha_{\rm CE}$). In population synthesis studies, some authors only allow a fraction $\alpha_{\rm th}$ of the available internal energy reservoir to contribute to the ejection, such that $\alpha_{\rm th}$ becomes a second efficiency parameter describing the problem.

3.2.2 Recombination Energy

Expanding and cooling plasma would inevitably recombine to neutral atoms, and then some atoms could form molecules. The energy released in these processes (the third term in Equation (3.7)) is typically referred to as recombination energy, $E_{\rm rec}$. Recombination energy was suggested a long time ago to be a potential driving mechanism for the ejection of ordinary planetary nebulae (Lucy 1967; Roxburgh 1967; Paczyński & Ziółkowski 1968). While the change in the overall storage of recombination energy during regular stellar evolution is usually not accounted for as a specific effect, the storage and release of recombination energy on a dynamical timescale takes place during pulsations of variable stars.

Recombination energy can be calculated by adding the appropriate ground state ionization and dissociation potentials for each ion and atom present (similar to Equation (3.7)), though dissociation to molecules is rarely included, and if it is done, then only H_2 is taken into account.

A rough estimate for a bulk value of the recombination energy from plasma to atoms can be made using the ionization potentials of hydrogen and helium, 13.6 eV per ion for H and 79.1 eV per ion for He. For a stellar mixture with hydrogen abundance X and helium abundance Y, the recombination energy is

$$u_{\rm rec} = 1.302 \times 10^{13} {\rm erg\ g^{-1}} \left(X + 5.809 \times \frac{Y}{4} \right). \tag{3.8}$$

For a typical stellar mixture with $X = 0.7$ and $Y = 0.28$, there are ten hydrogen ions per helium ion, resulting in an average ionization potential of 19.6 eV per ion, and a recombination energy $u_{\rm rec} \approx 1.5 \times 10^{13}$ erg g^{-1}. Using detailed stellar codes to take into account ions of other elements modifies this value slightly. A full calculation of this value for a giant star envelope, after a dredge-up has modified the envelope composition, gives $u_{\rm rec} \approx 1.6 \times 10^{13}$ erg g^{-1}. This energy would dominate the thermal energy for temperatures below $\sim 10^5$ K (since Boltzmann's constant $k_{\rm B}$ is 8.6×10^{-5} eV K^{-1}). This is about the same temperature below which helium recombination can start, and later hydrogen recombination.

Another way to evaluate the role of this energy is to compare it with the parameterized binding energy described by Equation (3.3). We can evaluate in which donor stars the recombination energy initially stored in the envelope exceeds the initial binding energy of the envelope:

$$R_{\rm d} > \frac{GM_{\rm d,i}}{\lambda u_{\rm rec}} \approx 120 R_\odot \times \frac{1}{\lambda}\frac{M_{\rm d,i}}{M_\odot}. \tag{3.9}$$

In such stars, if recombination is triggered, in principle it could remove the entire envelope.

The energy release from the two components, thermal and recombination, however, would dominate at different locations. The thermal term provides specific energy $u_{\rm th} \sim 3/2 k_B T$ per particle. It will store and release energy above a temperature of 10^5 K, i.e., deep within the star. The release of stored energy during recombination and molecule formation takes place below a temperature of 10^5 K. The location where plasma cools down and recombination takes place is a crucial factor in completing envelope removal. If recombination takes places above the "recombination radius" $R_{\rm rec} \approx 65 R_\odot \times (M_{\rm d,core}/M_\odot)$, and this energy is locally thermalized, the envelope can in principle be fully ejected through recombination outflows (see detailed discussion in Section 8.3).

A further plausible consequence of recombination is its effect on the opacity of any ejected matter, and so on the light curves of CEEs. This mechanism is similar to how type IIP supernova light curves are shaped. For suitable conditions this would lead to a "plateau" in the luminosity (Ivanova et al. 2013; see also Chapter 10).

However, whether the energy is thermalized (and hence can be used locally just as well as thermal energy) or is carried away by convection and then radiated away from the surface (and hence is lost) is the subject of ongoing discussion. What is clear is that its use cannot be the same for all cases. Ivanova 2018 have shown that the released energy will be thermalized locally if envelope expansion occurs on a dynamical timescale that is shorter than the thermal timescale and the envelope entropy prior to the CEE is $S/(k_B N_A) < 37$mol g^{-1} (for $X = 0.7$ and $Y = 0.28$). This occurs because under these conditions normal subsonic convection and radiation are unable to transport recombination energy away at the same rate as it is released. For CEEs which enter a slow spiral-in, recombination energy will be radiated away, but the same will be true for other involved energies including the released orbital energy. For example, this is likely to take place in systems with asymptotic giant branch donors (see e.g., Soker et al. 2018).

The efficiencies with which the recombination and thermal components of the envelope's internal energy can be used should not be assumed to be the same, just as $\alpha_{\rm CE}$ (parameterizing the use of energy from the orbital decay) is not expected to equal $\alpha_{\rm th}$. When using a simple parameterized energy formalism to estimate outcomes of CEEs, at least a third efficiency parameter, $\alpha_{\rm rec}$ for the recombination energy, ought to be included.

3.2.3 Calculating the Envelope Energy from Stellar Models

The energy of the envelope can be found using one-dimensional stellar models calculated with any stellar code using

$$E_{\rm env,i} = \int_{M_{\rm d}-M_{\rm d,env}}^{M_{\rm d}} \left(-\frac{Gm}{r} + u_{\rm th}(m) \right) dm. \tag{3.10}$$

The obtained energy is then usually converted into a λ value using the definition in Equation (3.3) and making the assumption that the binding energy is $E_{\rm bind} = -E_{\rm env,i}$:

$$\lambda = -\frac{GM_{\rm d,i}M_{\rm d,env}}{E_{\rm env,i}R_{\rm d}}. \tag{3.11}$$

Several groups have provided λ values for a range of donor masses, radii, core masses, and evolutionary stages; these are published either as tables or parameterized fits (recent examples include Loveridge et al. 2011; Wang et al. 2016). Modern population synthesis codes have switched to such calculated λ values rather than assuming a global constant value of λ (or combined $\alpha_{\rm CE}\lambda$) for all stars, as was done in the past.

Using values from stellar models, however, is not yet a recipe for a final answer. First of all, some groups choose not to include thermal energy in their binding (envelope) energies, which results in different values of λ for a given stellar structure (see Figure 3.1). When energy terms other than potential energy are included, it is not always clear if the authors include thermal energy only (as is most common), or internal energy including recombination. (By contrast, when groups provide multiple sets of λ values with different assumptions, work which makes use of those λ values has not always been careful to clarify which assumptions they are making. That is, population synthesis papers sometimes state "we adopt the λ values calculated by…" without specifying *which* lambda values.) Absolute energy levels must also be decided by the authors, who may make different choices. The zero energy level represents an additional degree of freedom that complicates comparisons.

It also has been known for a while that the definition of the boundary between the remnant core and the ejected envelope can dramatically affect the value of λ (Dewi & Tauris 2000; Tauris & Dewi 2001). The reason is that the closer one gets to the core, the larger are the specific potential energy and the internal energy; a hydrogen-burning shell often has a larger binding energy than the rest of the envelope. A common ad hoc assumption is that only the pre-CE convective envelope is ejected, placing the remnant core boundary at the base of that envelope. However, what exactly determines the final remnant mass is a major research question; various proposed definitions will be discussed in Section 3.2. Here we stress that the value of λ, even when extracted from the same stellar model, can be different by a factor

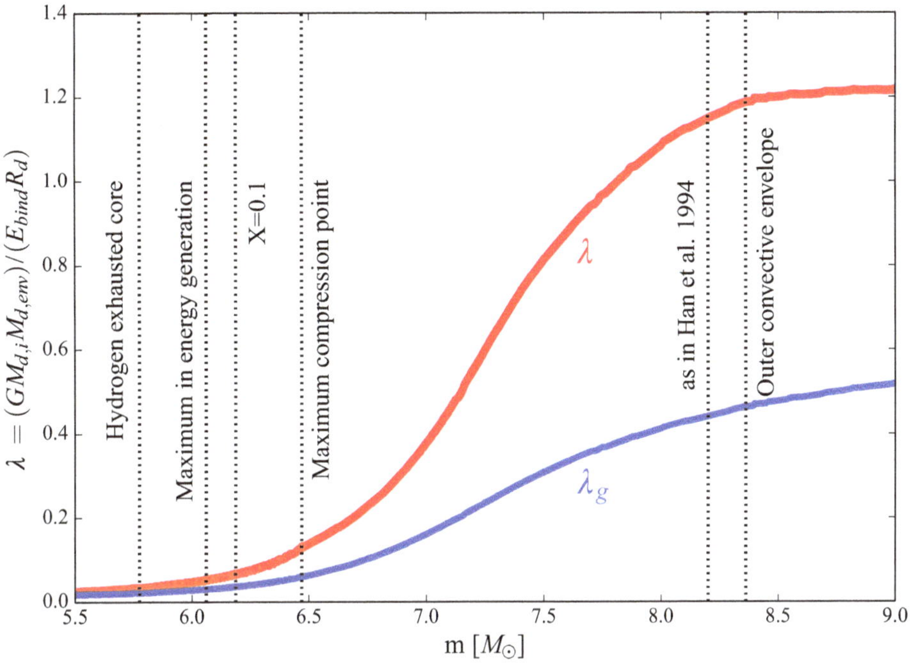

Figure 3.1. Structure parameter λ found using the energy of the envelope (Equation (3.11)) and λ_g found using only the potential energy of the envelope, as a function of mass. The example uses a $20M_\odot$ star when it has a radius $R = 750R_\odot$ ($Z = 0.02$). Dotted lines indicate different assumed final core masses (see Section 3.2).

of ~10–30 depending on the adopted core boundary definition (see Figure 3.1). Correspondingly, one has to be well aware that the outcomes of population synthesis calculations can vary dramatically depending on the core definition used (whether it is explicitly stated or not).

To further complicate the picture, the pre-plunge envelope structure can change significantly after the start of Roche-lobe overflow (RLOF). Pre-plunge mass transfer can continue over a thermal timescale, as we will discuss in Chapter 5, allowing plenty of time for the donor's thermal structure to change. Hence the range of λ values extracted from unperturbed single-star evolution might be very wrong.

3.2.4 Energetics of Steady-state Outflows

During a slow spiral-in the remaining common envelope, if unstable, may establish continuous outflows, e.g., powered by recombination (see Section 3.2.2 and Chapter 8). It has been argued that the condition to initiate continuous outflows capable of removing matter to infinity is similar to the Bernoulli equation, hence inclusive of an additional positive P/ρ term (Bisnovatyi-Kogan & Syunyaev 1972; Ivanova & Chaichenets 2011):

$$E_{\rm env,flow} = -\int_{M_{\rm d}-M{\rm d,\,env}}^{M_{\rm d}} \left(-\frac{Gm}{r} + u_{\rm th}(m) + \frac{P(m)}{\rho(m)} \right) dm. \tag{3.12}$$

The quantity $h = u + P/\rho$ is known as enthalpy, giving the condition above the name "enthalpy formalism." Since in Equation (3.12) the term $P/\rho > 0$, if continuous outflows are initiated, the envelope can be removed before its energy ($E_{\mathrm{env,i}}$; see Equation (3.10)) becomes positive. It is a subject of ongoing discussions of how exactly enthalpy works and whether P/ρ term indeed can provide an additional source of energy for envelope ejection. Below we discuss the implications that would result if it can work.

Unperturbed stellar models do not have a prior positive enthalpy throughout the entire envelope above the core. However, they could have "enthalpy positive" layers deep inside the envelope (Ivanova & Chaichenets 2011). The removal of the "pressure-cooker lid"—of the upper envelope—during the dynamical plunge-in helps to bring matter that was initially "enthalpy-unbound" to the surface.

At the same time, having a positive enthalpy does not guarantee the start of the outflows. To do so, an envelope with $E_{\mathrm{env,i}} < 0$ has to be first unstable with respect to adiabatic perturbation, when the pressure-weighted volume-averaged value of $\Gamma_1 < 4/3$. This condition was found to take place during a slow spiral-in (Ivanova et al. 2015).

We note that formally the enthalpy formalism may help to explain how low-mass companions might unbind stellar envelopes without requiring an apparent $\alpha_{\mathrm{CE}} > 1$. This is because it is derived without using *only* the envelope's energy budget, but indirectly may assume the use of the core's energy.

3.3 Extra Energy Sources

In addition to the conventional and intrinsically present mechanical and internal energies described above, additional sources of energy have been discussed as possibilities to help to unbind the envelope. None or all may be available, depending on the specifics of the particular CE system. Note that none of the energy sources or sinks described below can be simply inserted into the energy formalism, since they are not functions of the initial and final states.

3.3.1 Accretion Energy and Companion's Mass Gain

When the secondary moves through the envelope, some envelope material may accrete onto the companion, resulting in accretion luminosity. How effective can that luminosity be in affecting the CE energetic balance? A useful limiting estimate is the Eddington luminosity, which is $\sim 5 \times 10^{45} M_{\mathrm{comp}}/M_{\odot}\mathrm{erg\ yr}^{-1}$ for hydrogen-rich material. From Equation (3.11), the binding energy of the envelope is roughly

$$E_{\mathrm{bind}} \approx 3.8 \times 10^{48}\mathrm{erg}\frac{M_{\mathrm{d}}}{M_{\odot}}\frac{M_{\mathrm{d,env}}}{M_{\odot}}\frac{R_{\odot}}{R_{\mathrm{d}}}\frac{1}{\lambda}. \tag{3.13}$$

Let us consider that the plunge-in phase takes place on a dynamical timescale. The accretion energy, at the Eddington limit, generated during one dynamical time is

$$E_{\mathrm{accr}} \approx 5.7 \times 10^{41}\mathrm{erg}\frac{M_{\mathrm{comp}}}{M_{\odot}}\sqrt{\frac{(R_{\mathrm{d}}/R_{\odot})^3}{(M_{\mathrm{d}} + M_{\mathrm{comp}})/M_{\odot}}}. \tag{3.14}$$

Comparison of Equations (3.14) and (3.13) suggests that the accretion luminosity gained at the Eddington limit during a plunge is smaller than the binding energy for donors having $\lambda = 1$. It is substantially smaller if $\lambda \ll 1$ (this value is expected for massive evolved donors, see e.g., Podsiadlowski et al. 2003). The above suggests that accretion energy may play a role only for spiral-ins that take place on a very long timescale, such as self-regulated spiral-ins, if a slow spiral-in lasts very many dynamical times (as, e.g., considered in Ivanova 2002).

What is the expected accretion rate? A physical situation with the companion moving through a common envelope is similar to that of an object moving with some velocity through a uniform ambient medium. The latter is generally known as Bondi–Hoyle–Lyttleton accretion (Hoyle & Lyttleton 1939; Bondi & Hoyle 1944; see also the review in Edgar 2004). In Bondi–Hoyle–Lyttleton accretion, the gravity of the star focuses the gas as it travels past into a wake, which the star then accretes. Within the gravitational capture radius, the gas becomes supersonic and moves toward a freefall solution. This, as a whole, slows down the star's motion and increases its mass. In a common envelope, the velocity of the companion during most of the spiral-in is supersonic. Then the accretion rate can be approximated by the expression (see justifications in Edgar 2004):

$$\dot{M}_{\rm HL} = \frac{4\pi G^2 M_{\rm comp}^2 \rho_\infty}{v_\infty^3}. \tag{3.15}$$

Here ρ_∞ is the density of the common envelope, which in this approximation is assumed to be uniform, while v_∞ is the velocity of the companion relative to the background material. We introduce the average density of the donor, $\bar{\rho} = M_{\rm d}/(4/3\pi R_{\rm d}^3)$, and we approximate the relative velocity at infinity, relating it to the local sonic velocity in the common envelope as $v_\infty = \mathcal{M}_\infty c_\infty$. The sonic velocity, c_∞, can be estimated using the virial theorem:

$$c_\infty^2 \approx \Gamma_1 G M_{\rm d} \Big/ \left(3R_{\rm d}\right), \quad \Gamma_1 = \left(\frac{\partial \ln P}{\partial \ln \rho}\right)_{\rm ad}. \tag{3.16}$$

Here Γ_1 is the first adiabatic index. Bondi–Hoyle accretion is fairly similar to Hoyle–Lyttleton, with the exception that the sonic velocity is non-negligible, and the denominator becomes $(c_\infty^2 + v_\infty^2)^{3/2}$ (Edgar 2004). As a result, the Bondi–Hoyle accretion rate is smaller than the Hoyle–Lyttleton rate. It can be seen that the accretion rate during a common-envelope event, if predicted using the Hoyle–Lyttleton accretion rate expression, is:

$$\begin{aligned}\dot{M}_{\rm HL} &\approx \frac{3^{5/2} G^{1/2} M_{\rm comp}^2}{\Gamma_1^{3/2} M_{\rm d}^{1/2} R_{\rm d}^{3/2}} \\ \mathcal{M}_\infty^{-3}\frac{\rho_\infty}{\bar{\rho}} &\approx 1.4 \times 10^5 \frac{(M_{\rm comp}/M_\odot)^2}{(M_{\rm d}/M_\odot)^{1/2}(R_{\rm d}/R_\odot)^{3/2}} \frac{\rho_\infty}{\bar{\rho}} \mathcal{M}_\infty^{-3} \quad M_\odot\ {\rm yr}^{-1}.\end{aligned} \tag{3.17}$$

For example, a helium-star donor with a mass of $10M_\odot$, with an initial radius of $30R_\odot$, a companion of $1.4M_\odot$, and $\mathcal{M}_\infty = 10$, may be expected to accrete at $\dot{M}_{\rm HL} \approx 0.5(\rho_\infty/\bar{\rho})M_\odot\ {\rm yr}^{-1}$.

The actual accretion rate is usually Eddington-limited:

$$\dot{M}_{\rm Edd} = \frac{4\pi c}{\kappa} R_{\rm comp}. \tag{3.18}$$

Here κ is the mass absorption opacity coefficient near the companion's surface, and c is the speed of light. For the case of Thomson scattering opacity,

$$\dot{M}_{\rm Edd} \approx 2.2 \times 10^{-3} \frac{1}{1+X} \frac{R_{\rm comp}}{R_\odot} M_\odot\ {\rm yr}^{-1}. \tag{3.19}$$

As the companions are usually not giants, and so are a few solar radii or less, the actual mass gain during a common-envelope event is not expected to be huge for non-degenerate companions. The situation changes for degenerate companions. If the temperature in the neighborhood of an accreting compact object is sufficiently high, then thermal energy can be converted into neutrinos. The generated neutrinos remove a fraction of the potential energy released by accretion, without generating a radiation force for the removed fraction. The Eddington limit becomes inapplicable, and the compact object could accrete at well above the Eddington rate. This is known as hypercritical accretion.

For hypercritical accretion to occur, the photon radiation must be trapped in the flow while the flow travels outward to a point where the energy can be removed by neutrinos. Let us compare the infall velocity of the accreting material to the diffusion velocity of the radiation (Chevalier 1993). Assuming spherical inflow, the accretion velocity $v_{\rm acc}$ is related to the accretion rate onto the compact object $\dot{M}_{\rm acc}$ as:

$$\dot{M}_{\rm acc} = 4\pi r^2 \rho v_{\rm acc}. \tag{3.20}$$

Here ρ is the density of the accretion inflow at radius r. Consider now a photon diffusing out of the inflowing gas. After time $t_{\rm diff}$, the number of scatterings is $N = t_{\rm diff}c/\lambda_{\rm mfp}$, where $\lambda_{\rm mfp} = 1/(\rho\kappa)$ is the mean free path of a photon. For a random walk, the expectation value for the distance traveled is $r = \lambda_{\rm mfp}N^{1/2}$. So such a photon would diffuse to the radius r in a time $t_{\rm diff} = r^2/(\lambda_{\rm mfp}c)$. Hence the photon's diffusion velocity $v_{\rm diff}$ at the radius r is:

$$v_{\rm diff} = \frac{r}{t_{\rm diff}} = \frac{c}{\rho\kappa r}. \tag{3.21}$$

The radiation is trapped in the flow if $v_{\rm diff}$ is less than $v_{\rm acc}$. From Equations (3.18) and (3.20), this requires

$$\dot{M}_{\rm acc} > \frac{4\pi c r}{\kappa} = \dot{M}_{\rm Edd}\frac{r}{R_{\rm comp}}. \tag{3.22}$$

An additional condition is that the radius r inside which the photons are trapped should be larger than the radius inside which neutrinos are generated. A neutron star has a solid surface. Around it, neutrinos are formed within the accretion shock at the shock radius (Houck & Chevalier 1991):

$$R_{\mathrm{sh}} = 1.6 \times 10^{8}\left(\frac{\dot{M}_{\mathrm{acc}}}{M_{\odot}\, yr^{-1}}\right)^{-0.37} \mathrm{cm}. \quad (3.23)$$

This suggests the accretion rate on a neutron star should be as large as

$$\dot{M}_{\mathrm{hyper}} \geqslant 1.9 \times 10^{-4}\left(\frac{\kappa}{0.34\ \mathrm{cm}^2\ \mathrm{g}^{-1}}\right)^{-0.73} M_{\odot}\, yr^{-1} \approx 10^4\, M_{\mathrm{Edd}} \quad (3.24)$$

in order to become hypercritical.

As was shown above, the Hoyle–Littleton accretion rate can exceed this value for many massive giants (Fryer et al. 1996). Consequently, it has been argued that compact objects for which $r/R_{\mathrm{comp}} \gg 1$ would hypercritically accrete during a common-envelope event. In particular, this could turn a neutron star into a black hole and prevent the formation of close double neutron-star systems if the route involved the first-formed neutron star experiencing common-envelope evolution.

But what are the actual accretion rates? Modern high-resolution three-dimensional codes confirm that flows in uniform media, with negligible initial angular momentum, approach a steady-state solution at which the accretion rates are comparable to $\dot{M}_{\mathrm{HL}}$ (Blondin & Pope 2009; Blondin & Raymer 2012). However, when accretion occurs during a common-envelope spiral-in, the density of the medium is not uniform, and its angular momentum relative to the accretor is not negligible.

Indeed, full-fledged three-dimensional simulations of a common-envelope event have demonstrated that the effective accretion rate is significantly less than the rate based on a gravitational capture radius. For example, Ricker & Taam (2008) found that it is less than $\dot{M}_{\mathrm{HL}}$; they noticed that during a common-envelope event, the medium moves relatively slowly near the companion, and matter accretes subsonically, while a gravitational capture formalism requires supersonic accretion. At the same time, most of the gas in the vicinity of the companion is accelerated by its gravitational potential, gains angular momentum, and is slung far outwards, becoming now entirely unavailable for later accretion, see also Figure 3.2.

To study better the accretion onto the companion, one needs a much higher resolution in its vicinity. One way is to make a "wind-tunnel" style numerical experiment, as was performed by MacLeod & Ramirez-Ruiz (2015) and MacLeod et al. (2017); see also Figure 3.2. They also have found that the actual accretion rate is at least an order of magnitude less than that predicted by the Bondi–Hoyle–Lyttleton formalism (see Figure 3.3), but argued that the underlying physical reason for that is strong density gradients in a common envelope. Since the wind-tunnel simulations take place in an inertial reference frame, this conclusion is not at odds with Ricker & Taam (2008); but instead shows that part of the deficit in accretion

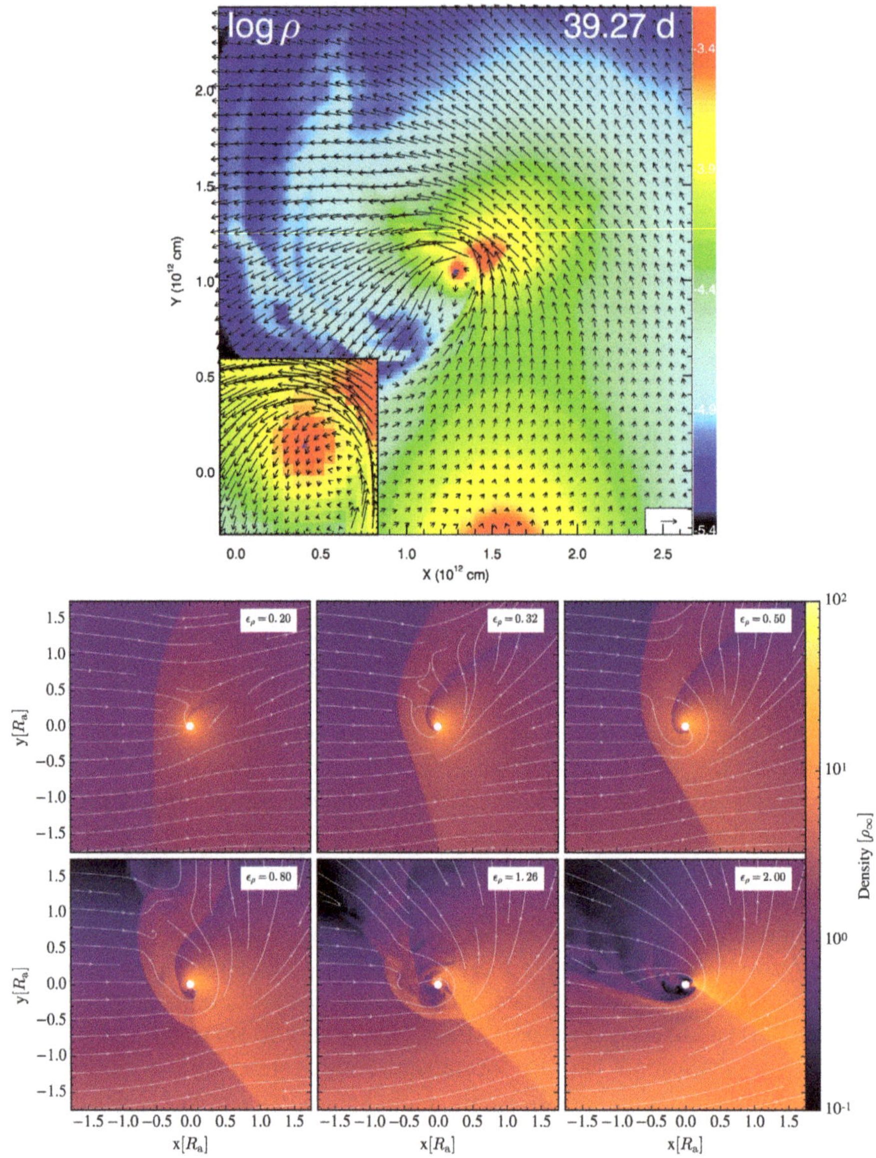

Figure 3.2. Flow structures (the density and the velocity field), in the orbital plane, near the companion. The top panel shows a full-fledged three-dimensional simulation of a common-envelope event. The gas density is shown in units of g cm^{-3}, with values indicated by the scale in the color bar. The arrow in the lower right-hand corner corresponds to 106 km s^{-1}. The blue arrow at the center indicates the companion's velocity. The red giant core lies just off the bottom of the plot. The inset shows the region immediately surrounding the companion, enlarged by a factor of three in each dimension. The bottom set of six panels shows wind-tunnel experiments. The color bar indicates the density of the medium, normalized to the density at infinity. These demonstrate that flow structures and net angular momentum develop more strongly as the density gradient increases (characterized by ε_ρ, the ratio of the gravitational focus radius R_a and the density scale height H_ρ). The top figure is reproduced from Ricker & Taam (2008) , and the bottom figure is reproduced from MacLeod et al. (2017)

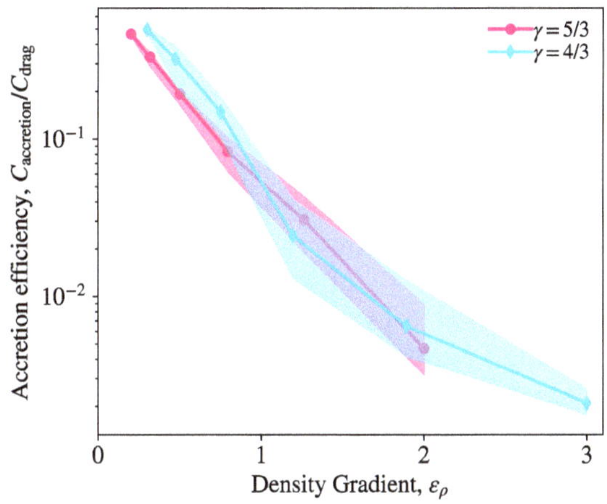

Figure 3.3. The ratio of the coefficient of accretion to the coefficient of drag (relative to gravitational-capture theory) as a function of density gradient. Density gradients are parameterized by the ratio of gravitational focus radius R_a to the density scale height H_ρ, $\varepsilon_\rho \equiv R_a/H_\rho$.Here the gravitational focus radius, also known as the Hoyle–Littleton accretion radius, is defined as $R_a = 2GM_{comp}/v_\infty^2$. While in uniform media both the coefficients of drag and accretion are of order unity, as significant density gradients are imposed on the flow, $C_{accretion} \ll C_{drag}$. The consequence for common-envelope phases is that objects gain much less mass than predicted by Bondi–Hoyle–Littleton theory alone. This figure was kindly generated by M. MacLeod specifically for this book, from the simulations published in MacLeod et al. (2017).

rate compared to the Hoyle–Lyttleton formalism comes from the density gradient. It is clear that the different numerical approaches agree in that the actual accretion rates are at least one order of magnitude smaller than the Hoyle–Littleton accretion rate, and the maximum amount of mass that a neutron star can accumulate during a common envelope event is $\sim 0.1 M_\odot$.

The accretion also affects the companion. If the companion is a non-degenerate star, it may expand dramatically upon accretion (Hjellming & Taam 1989). If the companion is a compact star, and the accretion rate is above the Eddington limit but less than hypercritical, then the material which is not accreted may drive a jet (see Section 3.4.3).

3.3.2 Nuclear Energy

New nuclear fusion in the perturbed deep layers of the donor star has been considered as a possible contributor to envelope ejection (Ivanova 2002; Podsiadlowski et al. 2010). This path is fairly uncommon and works only in very specific binaries, as many *ifs* are involved. Consider a binary that consists of a massive star at an advanced evolutionary stage, with two shells in which nuclear fusion is releasing energy, together with an unevolved low-mass companion. Now limit the case only to binaries which, according to the initial energy budget, are doomed to merge. *If* that binary does not merge immediately during the dynamical plunge-in phase, and transits into a self-regulated spiral-in regime, then during the latter phase, a non-compact companion can overfill its own Roche lobe. The start of

RLOF from the companion ends the canonical self-regulated spiral-in. The envelope keeps causing frictional drag on the companion, and the orbit continues to shrink. The reverse mass transfer will continue at an increasing rate. Since, in this case, the low-mass companion is unevolved, it mainly consists of hydrogen. A stream of low-entropy hydrogen-rich material penetrates deep down into the massive star and can reach the core (Ivanova et al. 2002). *If* this hydrogen-rich material reaches the He burning shell, then the ignition of this material can cause the shell to explode (Ivanova 2002; Podsiadlowski et al. 2010). The binding energy of the He shell in massive stars is a few times 10^{51} ergs, and that shell is not usually considered to be removed during a common envelope event. However, the released nuclear energy during explosive hydrogen burning can be larger than the binding energy, removing the He shell. The envelope outside the burning shells, which is typically considered to be the common envelope, has a binding energy substantially smaller than that of the He shell, and it will be plowed away by the He shell ejecta.

The outcome of the nuclear-powered explosive common-envelope event is a compact binary consisting of the (stripped) core of the giant and the (stripped) post-mass-transfer remnant of the low-mass companion. A direct application of this exotic channel is that companions may eject a common envelope despite having a mass far smaller than would be required by the energy formalism. This channel even favors lower-mass companions, since they have lower entropies. In particular, this channel may help with the formation of short-period low-mass X-ray binaries containing black holes, for which the classic common-envelope channel is particularly challenging when analyzed with the energy formalism and realistic binding-energy parameters (λ). Besides, the mass transfer provides the core of the giant not only with hydrogen but also with angular momentum, hence potentially producing a fast-rotating core that has been stripped of both hydrogen and helium (Podsiadlowski et al. 2010). The post-common-envelope remnant of the donor could then produce a long-duration γ-ray burst from a core stripped of both hydrogen and helium, consistent with the observed connection between long γ-ray bursts and type Ic SNe.

3.4 Ways in Which the Energy Reservoirs May Be Used

In the previous section we considered the sources of energy available for envelope ejection. Here we consider some processes through which the energy can be used to accomplish the ejection. In common-envelope theory this question is relatively unexplored. However, several processes have been considered in the literature, though erroneously they are sometimes treated as additional sources of energy. We discuss these processes here.

3.4.1 Tidal Heating

Tides affect both the stellar orbits and the rotation of the envelope, using orbital energy as a reservoir. Orbital energy release can be tidally mediated either directly, by pushing on the envelope, or indirectly, by frictionally heating the envelope, producing internal energy that is then used to eject it. The tidal circularization and synchronization timescales are similar and can be estimated for circular orbits using Equation (5.4). Typically they are very long when the envelope's rotation is nearly synchronized with the orbit. The timescales are, however, strong functions of eccentricity and relative degree of synchronization (Hut 1981), and they may both be relevant to the use of orbital energy release. To date the relative importance of the two processes has not been carefully investigated. A resolution of this question may help in understanding the efficiency of orbital energy release in envelope ejection. For example, if indirect usage (tidal heating) has a shorter timescale than direct usage (tidal acceleration), and if use of internal energy has a low efficiency, then the overall usage of orbital energy would also have a low efficiency.

3.4.2 Magnetic Fields

The role of magnetic fields in common-envelope events is one more piece of physics that is far from fully understood. During the spiral-in the envelope is expected to spin up, providing an opportunity for an effective magnetic dynamo to operate. A dynamo will not generate new energy but will redistribute energy already present in the system. Any increase in magnetic energy comes from decreases in other parts of the energy budget and, in this respect, it acts as a sink of energy. The initial magnetic field can be considered as an additional energy reservoir, but typical stellar magnetic fields do not represent a significant reservoir of energy (compared to the overall energy budget, i.e., to the envelope's potential energy and internal energy).

It has been argued that the generation of magnetic fields during common-envelope evolution could explain magnetic white dwarfs observed to have fields of 10^6–10^9 G (Tout et al. 2010), either by a large-scale dynamo in the envelope, or by a dynamo generated in the accretion disk (Nordhaus et al. 2011). However, it has also been argued that the magnetic field of such a white dwarf, if created, may quickly decay after the end of the common-envelope event (Potter & Tout 2010).

The first self-consistent, three-dimensional hydrodynamic common-envelope simulation that included magnetic fields demonstrated that the fields are strongly amplified in the accretion stream around the companion (Ohlmann et al. 2016). The generated magnetic field reached strengths of 10–100 kG throughout the envelope. However, even at this strength, the magnetic fields were found to be dynamically irrelevant for the ejection of the common envelope itself. Nonetheless such fields may alter the outflow geometry, possibly shaping post-CE nebulae in collimated bipolar outflows as was proposed by Nordhaus et al. (2007). If a circumbinary disk forms during the slow spiral-in phase, it can amplify the magnetic field enough to collimate later outflows (García-Segura et al. 2020).

3.4.3 Jets

Accretion energy release might be able to help envelope ejection in ways other than via heating from radiative accretion luminosity. In some circumstances accretion generates jets, i.e., fast, collimated outflows. A jet is another way in which energy liberated by accretion may couple to the envelope, not an intrinsically new source of energy.

Soker (2004) argued that formation of jets should be commonplace when the inspiraling companion is a white dwarf or a neutron star, and that the formation of some type of jet by a non-degenerate star cannot be excluded. Jets may either blow "hot bubbles" within the envelope, where the jet shocks the common envelope material, or they may mechanically "push" some common envelope matter outwards. In either case, jets might cause additional mass ejection from the common envelope. Whether or not the jets directly eject material, if the jets decrease the density of the material in the vicinity of the companion then they may slow down the rate of the spiral-in.

The physics of jet formation is not well-understood in general, and jets inside common envelopes are even further from being understood. In addition, there are no direct observational constraints on the existence of jets during common-envelope events, or on their opening angles, or their specific kinetic energies. However, if jets do form, then depending on their assumed initial jet velocity and opening angle, three-dimensional simulations find that strong polar outflows could form, ejecting the outer parts of the envelope (Moreno Méndez et al. 2017; Chamandy et al. 2018; Shiber et al. 2019; López-Cámara et al. 2019). The outcomes so far, however, are such that jets do not help to unbind the entire envelope, and the systems at the end of the simulations are too wide as compared with observed post-CE binaries (Shiber et al. 2019).

3.5 Energy Losses: Radiation

During the entire common-envelope event, the expanding envelope continues to lose energy from its surface. For a rough estimate of the rate of energy losses one can use $L/L_\odot = (R_{\rm CE}/R_\odot)^2(T_{\rm eff}/T_{\rm eff,\odot})^4$. The donor can continue to produce nuclear luminosity $L_{\rm nuc}$, which will be either the same as at the donor's pre-CE state, or smaller if the shell sources are inhibited. The radiation losses are thus a net sink if the new surface luminosity exceeds the initial nuclear energy luminosity, $L_{\rm CE} > L_{\rm nuc}$. As we discuss below, this is usually the case during a CEE. The initial balance of the surface luminosity and the nuclear luminosity explains why initial nuclear energy production is not taken into account in the energy balance equation.

During a plunge-in, the non-yet-ejected envelope usually does not expand by more than a factor of ten. At the same time, the temperature of the layer from which the photons escape typically does not drop below ~3000 K. This allows us to estimate the maximum luminosity during the plunge-in as $L \approx 10L_\odot(R_{\rm d,i}/R_\odot)^2$. A plunge-in takes place on a dynamical timescale, limiting the total radiative losses to

$$E_{\rm rad} \approx 6 \times 10^{37} {\rm erg}\ (R_{\rm d,i}/R_\odot)^{3.5}(M_{\rm d,i}/M_\odot)^{-1/2}. \tag{3.25}$$

Comparing this result with the binding energy, $E_{\rm bind} \approx 4 \times 10^{48} {\rm erg}\ (R_{\rm d,i}/R_\odot)^{-1}$ $(M_{\rm d,i}/M_\odot)^2$, indicates that for prompt envelope ejections that take place on a dynamical timescale, radiative losses are important only for stars which are larger than $R \approx 250R_\odot(M_{\rm d,i}/M_\odot)^{0.55}$. Therefore, radiative losses may affect AGB or supergiant donors during the dynamical stage of a CEE, but should be not significant for most donors.

On the other hand, the self-regulated spiral-in lasts for many orbits, approaching the thermal timescale of the expanded envelope. Radiative losses during the self-regulated spiral-in cannot be ignored, although predicting them without calculating a CEE in its entirety is challenging, if not impossible. This suggests that the energy conservation approach is not useful for predicting the outcome of any CEE which enters a self-regulated stage.

3.6 The Complete Energy Budget

Simple use of the energy formalism, which assumes a single $\alpha_{\rm CE}$ for all common-envelope events, is undoubtedly convenient. However, it is increasingly clear that this assumption has little, if any, relation to the physics, and such predictions may well be misleading. Assuming a single, fixed value of $\alpha_{\rm CE}$ is not a good description of our understanding of the physics. Additionally, since λ is known to be sensitive to evolutionary stage, the use of a constant $\alpha_{\rm CE}\lambda$ value, which implies $\alpha_{\rm CE} \propto \lambda^{-1}$, is not supported by our present understanding. Both choices seem likely to introduce unphysical systematic features into predictions which adopt them.

From numerical studies of common-envelope events and their energy budgets, even binary systems with similar low-mass giant donors would not evolve via a CEE with the same efficiency (e.g., Nandez & Ivanova 2016; in which effective efficiencies are found to change by roughly a factor of two). From analytical expectations, at least three different "efficiencies" are important, as described above.

Observationally, the properties of post-CE systems which are inferred to have had pre-CE RGB and AGB donor stars suggest that CE events with AGB donors have proceeded with a systematically lower efficiency of energy re-use than with RGB donors (Iaconi & De Marco 2019), although there is overlap between the inferred $\alpha_{\rm CE}$ values for systems in the two groups. Note that the apparent inefficiency for AGB donors could indicate that such donors mainly experience self-regulated CEEs during which most of the released energy is radiated away, or that the binary shrank during very efficient wind-Roche-lobe overflow even without entering a CE phase (Chen et al. 2020).

The emerging approach is to write an energy conservation equation that is as complete as possible while decomposing it into well-distinguished energy components (e.g., Nandez & Ivanova 2016; Ivanova & Nandez 2016; Chamandy et al. 2019). The goal is that, in the future, it might be possible to constrain each of the components separately, using appropriate numerical methods, and then wholly forgo $\alpha_{\rm CE}$. The evolution of different energy components may help in building a proper algorithm for calculating CE in one-dimensional stellar codes (see the discussion and initial steps in Ivanova & Nandez 2016). Since all the constraints will be determined using different numerical approaches, as a community we have arrived at the end of purely analytical studies of common envelope events. The future belongs to numerical methods, a review of which is given in Chapter 4. Here we will briefly describe the energy components.

The most generic form of energy conservation for the initial and the final states can be written as

$$U^{\rm ini} + \Omega^{\rm ini} + K^{\rm ini} + E_{\rm extra} = U^{\rm fin} + \Omega^{\rm fin} + K^{\rm fin} + E_{\rm rad}, \tag{3.26}$$

U stands for internal energy, Ω stands for gravitational potential energy, K stands for bulk kinetic energy, and indexes ini and fin indicate the initial and the final states of the common envelope system. $E_{\rm extra}$ combines any of the extra sources of energy described above—accretion energy (potentially mediated by jets), initial magnetic fields, energy from nuclear reactions, etc. $E_{\rm rad}$ represents the energy sink due to radiative energy losses from the surface of the common envelope.

Below we will consider first the components of the internal, potential, and bulk kinetic energies which belong to the participating stars at the start and at the end of the common-envelope event, and then separately, the same terms for the ejected material.

3.6.1 The Internal Energy

At each moment, the internal energy is composed of not only thermal and recombination energies of the envelope, but also of the internal energies of the donor's core and of the companion:

$$U = U_{\rm env,th} + U_{\rm env,rec} + U_{\rm core} + U_{\rm comp}. \tag{3.27}$$

Our understanding of how the internal energies of the core $U_{\rm core}$ and of the companion $U_{\rm comp}$ can affect CE evolution is still in a state of infancy. Changes of $U_{\rm core}$ and $U_{\rm comp}$ during a CEE are not taken into account by the energy budget equations which have been used to date. Also, the unchanging internal energy of the donor's core is implicitly adopted by currently existing three-dimensional hydrodynamic codes. These codes model the core as a special particle that interacts only gravitationally but otherwise has no notion of its own internal energy. As one-dimensional codes have shown, the core might be a reservoir of thermal energy. For example, it can expand during a CEE as a reaction to the envelope's removal or expansion (Deloye & Taam 2010; Ivanova 2011). In some three-dimensional codes, a common envelope event is modeled with a companion that is allowed to have internal structure. In this approach, the companion's internal energy is implicitly taken into account in the outcomes. Sometimes the companion could be destroyed during a merger, which is energetically dramatically different from the case in which the structure of the companion is neglected (Gaburov et al. 2008).

3.6.2 The Potential Energy

The potential energy is composed partly of familiar and partly of less commonly considered terms:

$$\Omega = \Omega_{\rm env,selfgr} + \Omega_{\rm env-core} + \Omega_{\rm env-comp} + \Omega_{\rm core-comp} + \Omega_{\rm core,selfgr} + \Omega_{\rm comp,selfgr}. \tag{3.28}$$

The potential energy of the envelope includes its self-gravity $\Omega_{\rm env,selfgr}$ and its gravitational interaction with the core $\Omega_{\rm env-core}$. The potential energy due to gravitational interactions between the donor star's components and the companion are denoted by $\Omega_{\rm env-comp}$ and $\Omega_{\rm core-comp}$. However, two terms are usually forgotten: the self-gravity of the core, $\Omega_{\rm core,selfgr}$, and the self-gravity of the companion, $\Omega_{\rm comp,selfgr}$. As with the internal energies of the core and companion, the last two terms are not usually considered to change throughout the CE evolution; this assumption may well be wrong, especially for the cores of massive donor stars.

Let us analyze the terms of this equation from the point of view of the energy formalism. For the initial state, the first two terms are accounted for by the gravitational potential energy of the envelope:

$$\Omega_{\rm env} = \Omega_{\rm env,selfgr} + \Omega_{\rm env-core}. \tag{3.29}$$

$\Omega_{\rm env}$ is then used as a component of the binding energy of the envelope.

The next two terms in Equation (3.28), $\Omega_{\rm env-comp}$ and $\Omega_{\rm core-comp}$, are accounted for in the energy formalism by the orbital energy, both for the initial state (prior to the CEE, at the start of RLOF), and for the final state (after the CEE). Note, however, that, at the start of a CEE, the sum of these two terms is not equal to the simple expression used in the energy formalism, $-GM_{\rm d,i}M_{\rm comp}/a^{\rm ini}$.

Finally, the last two terms in Equation (3.28), $\Omega_{\rm core,selfgr}$ and $\Omega_{\rm comp,selfgr}$, are usually ignored. These terms are treated in a way similar to the internal energies of the core and the companion: the changes during a CEE are either ignored or, in rare three-dimensional simulations, are taken into account implicitly for the companion. The change in $\Omega_{\rm comp,selfgr}$ during a CEE, and the consequence of that, has not so far been isolated in any published analysis.

During the plunge-in phase, splitting the potential energy into components is even more complicated; this is discussed in Section 6.4.

3.6.3 The Bulk Kinetic Energy

The bulk kinetic energy consists of the kinetic energies of the donor's core, the companion, and the donor's envelope:

$$K_{\rm tot} = \frac{1}{2}M_{\rm core}v_{\rm core}^2 + \frac{1}{2}M_{\rm comp}v_{\rm comp}^2 + K_{\rm env}. \tag{3.30}$$

For the initial state, it often makes sense to split the initial kinetic energy of the envelope into the kinetic energy of the orbital motion $K_{\rm env,orb}^{\rm ini}$ and the kinetic energy of the motion relative to the orbital motion $K_{\rm env,rel}^{\rm ini}$. The latter is zero if, at RLOF, the donor is fully synchronous with the orbital motion.

As with the term $\Omega_{\rm env-comp}$, since the envelope is not a point mass, the simplification used in the energy formalism is not valid even for initial states in which the envelope is fully synchronized with the orbit:

$$K^{\rm ini}_{\rm env,orb} + K^{\rm ini}_{\rm env,rel} + \frac{1}{2}M_{\rm core}\left(v^{\rm ini}_{\rm core}\right)^2 + \frac{1}{2}M_{\rm comp}\left(v^{\rm ini}_{\rm comp}\right)^2 \neq \frac{GM_{\rm d,i}M_{\rm comp}}{2a^{\rm ini}}. \tag{3.31}$$

During the plunge-in phase, the orbit cannot be described using simple Keplerian terms (see Chapter 6). After the plunge-in phase, the energy formalism considers the kinetic energy to be composed only of the orbital kinetic energy of the newly formed binary, $GM_{\rm core}M_{\rm comp}/(2a^{\rm fin})$. The true kinetic energy of the new binary configuration $\frac{1}{2}M_{\rm core}(v^{\rm fin}_{\rm core})^2 + \frac{1}{2}M_{\rm comp}(v^{\rm fin}_{\rm comp})^2$, includes the kinetic energy of the center of mass motion, $K_{\rm bin,COM}$:

$$\frac{1}{2}M_{\rm core}\left(v^{\rm fin}_{\rm core}\right)^2 + \frac{1}{2}M_{\rm comp}\left(v^{\rm fin}_{\rm comp}\right)^2 = \frac{GM_{\rm core}M_{\rm comp}}{2a^{\rm fin}} + K_{\rm bin,COM}. \tag{3.32}$$

One potential reservoir of bulk kinetic energy which we have not described above is the energy stored in convective motions. In low-mass pre-CE donor stars, the convective velocities are typically expected to be below ten percent of the sound speed, so the kinetic energy stored in convective motions should be less than one percent of a dynamically-significant energy budget. For massive donor stars, in which convective velocities are typically larger fractions of the sound speed, it is less clear that these motions should be neglected.

3.6.4 The Energy of the Ejecta

In the final state, the standard energy formalism does not account for the components of the total energy of the ejected material $E_{\rm ej}$: its internal energy $U_{\rm ej}$, potential energy $\Omega_{\rm ej}$, and kinetic energy $K_{\rm ej}$. They all are taken to be zero. In three-dimensional simulations performed to date, these energy terms have been found to be significant; $K_{\rm ej}$ is up to about 40 percent of the final orbital energy, which can be comparable to the initial binding energy (for more details see Chapter 9).

To summarize, the final kinetic energy is

$$K^{\rm fin}_{\rm tot} = \frac{GM_{\rm core}M_{\rm comp}}{2a^{\rm fin}} + K_{\rm bin,COM} + K_{\rm ej}. \tag{3.33}$$

Only the first term on the right-hand side is used in the energy formalism.

3.7 A Brief Guide to the Energy Components

For a proper understanding of the physics of common-envelope events, it is quite useful to characterize an event in terms of the energy components and timescales involved. In table 3.7, we list all the energy components discussed previously while briefly commenting on whether our understanding of that term has any known problem.

Energy Component	What It Is	Status, Problems, Comments, Timescales
$U_{\mathrm{env,th}}$	Thermal energy of the common envelope	Problem: there is no validated efficiency of conversion of thermal energy into mechanical energy. The initial value can be calculated from one-dimensional stellar structure if there would be a proven method to determine which part of the envelope is ejected and which is not.
$U_{\mathrm{env,rec}}$	Recombination energy of the common envelope	This term represents a reservoir of energy which, if released, may enable complete envelope ejection, at least in some CE cases. The total amount available can be found from the initial structure, but the efficiency of use is not established.
U_{env}	Internal energy of the common envelope	This term consists of the thermal energy and of the recombination energy of the common envelope. See the problems described above for those terms.
U_{ej}	Thermal energy of the ejected envelope	This energy term is currently ignored explicitly by the energy formalism, while implicitly it is included via $\alpha_{\mathrm{CE}} < 1$. This term can be non-negligible.
U_{core}	Internal energy of the remnant	This term is currently treated as a non-changing quantity throughout a CEE. However, the core can expand upon envelope removal and change the overall energy budget. Timescale: dynamical to thermal timescale of the core.
U_{comp}	Internal energy of the companion	This term is often considered to be a non-changing quantity throughout a CEE. However, the companion's tidal destruction can modify the overall energy budget. Timescale: dynamical to thermal timescale of the companion.
K_{ej}	Final kinetic energy of the ejecta	This term is currently ignored explicitly by the energy formalism, while implicitly it is included via $\alpha_{\mathrm{CE}} < 1$. According to current numerical simulations, this term is non-negligible.
$K_{\mathrm{bin,COM}}$	Kinetic energy of the center of mass motion of the post-CE binary	This term determines how far the post-CE binary can move away from the ejected common envelope, and may be tested with observed planetary nebulae. The origin of this energy is due to asymmetrically ejected material. This term is smaller than the other terms in the total energy budget, but is linked with the shape of the post-CE nebula.

$K_{\rm env}^{\rm ini}$	Initial kinetic energy of the envelope	This energy term is an intrinsically three-dimensional quantity and cannot be reliably approximated by treating the envelope as a point mass or as spherically symmetric.
$\Omega_{\rm env}$	Gravitational potential energy of the envelope	This energy term is an intrinsically three-dimensional quantity and cannot be reliably approximated by treating the envelope as a point mass or as spherically symmetric when the donor star is near to filling, or overflowing, its Roche lobe. The canonical energy formalism uses a simplified point-mass approximation for the initial binding energy of the envelope.
$\Omega_{\rm env-comp}$	Potential energy between the envelope and the companion	This energy term is an intrinsically three-dimensional quantity and cannot be reliably approximated using a point-mass or spherically symmetric approximation when the donor star is approaching its Roche lobe or overflows it. The canonical energy formalism uses the simplified point-mass approximation for the initial orbital energy of the envelope.
$\Omega_{\rm core,selfgr}$ and $\Omega_{\rm comp,selfgr}$	Self-gravity of the core and the companion	These energy terms are currently treated as being constant throughout a common-envelope event. As with the thermal energies of the remnant and the core, in reality these energies might change during the event and so affect the energy budget.
$E_{\rm env,flow}$	Enthalpy formalism estimate of envelope energy	This term has been used as an alternative condition to remove an unstable common envelope through quasi-stationary outflows. Timescale: thermal timescale of the common envelope.
$E_{\rm acc}$	Energy released due to accretion onto the companion	The accretion efficiency depends on the local envelope's density, entropy, and angular momentum, which are hard to treat self-consistently for accretion calculations. Nonetheless this source of energy is typically considered negligible in most cases. Timescale: time of the actual spiral-in; i.e., dynamical for prompt CEEs and thermal for self-regulated spiral-ins.

(*Continued*)

(*Continued*)

Energy Component	What It Is	Status, Problems, Comments, Timescales
E_{jet}	The energy of jets produced by the companion	This is a different way for accretion to potentially affect the common envelope. If the energy input is as direct kinetic energy, then simulations find the use of this accretion energy to be more efficient than if the input is radiative heating. The formation of jets is not well understood. Timescale: as for accretion.
E_{nuc}	Additional explosive nuclear energy	Even during a slow spiral-in, nuclear reactions may continue as before the common-envelope event, or they may be quenched due to core expansion and cooling. This term, however, considers a plausible dynamical-timescale nuclear energy release due to the penetration of companion material into hot burning layers of the donor. The released energy by far exceeds the orbital energy, making the energy formalism inappropriate. Explosive nuclear burning itself is very fast, but it can take place at any time during a dynamical plunge or during a slow spiral-in.
E_{rad}	Radiative energy losses	Amount of energy lost by radiation from the photosphere of the expanding common envelope. It is comparable to the overall energy during a slow spiral-in, making the energy formalism inappropriate. Timescale: thermal timescale of the expanded envelope.

References

Bisnovatyi-Kogan, G. S., & Syunyaev, R. A. 1972, SvA, 15, 697

Bisnovatyi-Kogan, G. S., & Zeldovich, Y. B. 1967, SvA, 10, 959

Blondin, J. M., & Pope, T. C. 2009, ApJ, 700, 95

Blondin, J. M., & Raymer, E. 2012, ApJ, 752, 30

Bondi, H., & Hoyle, F. 1944, MNRAS, 104, 273

Chamandy, L., Frank, A., Blackman, E. G., et al. 2018, MNRAS, 480, 1898

Chamandy, L., Tu, Y., Blackman, E. G., et al. 2019, MNRAS, 486, 1070

Chen, Z., Ivanova, N., & Carroll-Nellenback, J. 2020, ApJ, 892, 110

Chevalier, R. A. 1993, ApJL, 411, L33

de Kool, M. 1990, ApJ, 358, 189

Deloye, C. J., & Taam, R. E. 2010, ApJL, 719, L28

Dewi, J. D. M., & Tauris, T. M. 2000, A&A, 360, 1043
Edgar, R. 2004, NewAR, 48, 843
Fryer, C. L., Benz, W., & Herant, M. 1996, ApJ, 460, 801
Gaburov, E., Lombardi, J. C., & Portegies Zwart, S. 2008, MNRAS, 383, L5
García-Segura, G., Taam, R. E., & Ricker, P. M. 2020, ApJ, 893, 150
Han, Z., Podsiadlowski, P., & Eggleton, P. P. 1994, MNRAS, 270, 121
Hjellming, M. S., & Taam, R. E. 1989, BAAS, 21, 1081
Houck, J. C., & Chevalier, R. A. 1991, ApJ, 376, 234
Hoyle, F., & Lyttleton, R. A. 1939, PCPS, 35, 405
Hut, P. 1981, A&A, 99, 126
Iaconi, R., & De Marco, O. 2019, MNRAS, 490, 2550
Iben, I. Jr, & Tutukov, A. V. 1984, ApJS, 54, 335
Ivanova, N. 2002, DPhil thesis, Balliol College, Oxford
Ivanova, N. 2011, ApJ, 730, 76
Ivanova, N. 2018, ApJL, 858, L24
Ivanova, N., & Chaichenets, S. 2011, ApJL, 731, L36
Ivanova, N., Justham, S., Avendano Nandez, J. L., & Lombardi, J. C. 2013, Sci, 339, 433
Ivanova, N., Justham, S., & Podsiadlowski, P. 2015, MNRAS, 447, 2181
Ivanova, N., & Nandez, J. L. A. 2016, MNRAS, 462, 362
Ivanova, N., Podsiadlowski, P., & Spruit, H. 2002, MNRAS, 334, 819
Livio, M., & Soker, N. 1988, ApJ, 329, 764
López-Cámara, D., De Colle, F., & Moreno Méndez, E. 2019, MNRAS, 482, 3646
Loveridge, A. J., van der Sluys, M. V., & Kalogera, V. 2011, ApJ, 743, 49
Lucy, L. B. 1967, AJ, 72, 813
MacLeod, M., Antoni, A., Murguia-Berthier, A., Macias, P., & Ramirez-Ruiz, E. 2017, ApJ, 838, 56
MacLeod, M., & Ramirez-Ruiz, E. 2015, ApJ, 798, L19
Moreno Méndez, E., López-Cámara, D., & De Colle, F. 2017, MNRAS, 470, 2929
Nandez, J. L. A., & Ivanova, N. 2016, MNRAS, 460, 3992
Nordhaus, J., Blackman, E. G., & Frank, A. 2007, MNRAS, 376, 599
Nordhaus, J., Wellons, S., Spiegel, D. S., Metzger, B. D., & Blackman, E. G. 2011, PNAS, 108, 3135
Ohlmann, S. T., Röpke, F. K., Pakmor, R., Springel, V., & Müller, E. 2016, MNRAS, 462, L121
Paczyński, B., & Ziółkowski, J. 1968, AcA, 18, 255
Podsiadlowski, P., Ivanova, N., Justham, S., & Rappaport, S. 2010, MNRAS, 406, 840
Podsiadlowski, P., Rappaport, S., & Han, Z. 2003, MNRAS, 341, 385
Potter, A. T., & Tout, C. A. 2010, MNRAS, 402, 1072
Ricker, P. M., & Taam, R. E. 2008, ApJL, 672, L41
Roxburgh, I. W. 1967, Natur, 215, 838
Shiber, S., Iaconi, R., De Marco, O., & Soker, N. 2019, MNRAS, 488, 5615
Soker, N. 2004, NewA, 9, 399
Soker, N., Grichener, A., & Sabach, E. 2018, ApJL, 863, L14
Tauris, T. M., & Dewi, J. D. M. 2001, A&A, 369, 170
Tout, C. A., Wickramasinghe, D. T., Liebert, J., Ferrario, L., & Pringle, J. E. 2010, in AIP Conf. Ser. 1314, International Conference on Binaries, ed. V. Kalogera, & M. van der Sluys (Melville, NY: AIP), 190

Tutukov, A., & Yungelson, L. 1979, in IAU Symp. 83, Mass Loss and Evolution of O-Type Stars, ed. P. S. Conti, & C. W. H. De Loore (Berlin: Springer), 401
van den Heuvel, E. P. J. 1976, in IAU Symp. 73, Structure and Evolution of Close Binary Systems, ed. P. Eggleton, S. Mitton, & J. Whelan (Berlin: Springer), 35
Wang, C., Jia, K., & Li, X.-D. 2016, RAA, 16, 126
Webbink, R. F. 1984, ApJ, 277, 355

Natalia Ivanova, Stephen Justham and Paul Ricker

Chapter 4

The Codes That Do the Job

The common envelope problem involves a number of physical processes interacting over a large range of length and timescales. Because of its complexity, numerical simulations have been essential in theoretical studies of common-envelope evolution (CEE). These tools have shaped our understanding of the behavior and outcomes of CEE. In this Chapter we discuss how the physics of CEE is numerically modeled using 3D simulation codes, and the limitations of the numerical methods and approximations in use.

4.1 Physics of Common-envelope Evolution

Here we consider first the range of mechanisms that must be addressed in simulations.

4.1.1 Equations of Hydrodynamics

Under most circumstances the matter in a star consists of atoms or molecules that are not strongly interacting; in other words, they can be treated as a gas. Statistically, a gas is described using a single-particle distribution function $f(\mathbf{x}, \mathbf{v}, t)$, which gives the probability of finding a particle with position $\mathbf{x}$ and velocity $\mathbf{v}$ at time t. Given an appropriate coarse-graining scale $L \gg \ell$, where ℓ is the typical interparticle separation, we can define local velocity moments

$$M_{k,i}(\mathbf{x}, t) \equiv \frac{1}{L^3} \int d^3x' \int d^3v' \, (v_i')^k f(\mathbf{x}', \mathbf{v}', t), \tag{4.1}$$

where the spatial integral is taken over a cube of side L centered on $\mathbf{x}$. If in addition L is much larger than the collisional mean free path λ, we can accurately describe the gas as a fluid using the first three velocity moments,

$$\text{number density } n \equiv M_0 \tag{4.2}$$

doi:10.1088/2514-3433/abb6f0ch4

$$\text{velocity } \mathbf{v} \equiv [M_{1,1}, M_{1,2}, M_{1,3}]/n \tag{4.3}$$

$$\text{specific internal energy } u \equiv \frac{1}{2}\sum_{i=1}^{3}\left(\frac{M_{2,i}}{n} - v_i^2\right), \tag{4.4}$$

where we have omitted the second subscript for $k = 0$ and assumed $v \equiv |\mathbf{v}| \ll c$. If the particles have mass m, the mass density $\rho = mn$. These quantities evolve according to the Navier–Stokes equations (in conservation form)

$$\frac{\partial \rho}{\partial t} + \nabla \cdot (\rho \mathbf{v}) = 0 \tag{4.5}$$

$$\frac{\partial \rho \mathbf{v}}{\partial t} + \nabla \cdot (\rho \mathbf{v}\mathbf{v}) = -\nabla P - \rho \nabla \phi + \nabla \cdot \boldsymbol{\pi} \tag{4.6}$$

$$\frac{\partial \rho E}{\partial t} + \nabla \cdot [(\rho E + P)\mathbf{v}] = -\rho \mathbf{v} \cdot \nabla \phi + \nabla \cdot (\boldsymbol{\pi} \cdot \mathbf{v}) - \nabla \cdot \mathbf{F} + \rho \dot{q}, \tag{4.7}$$

where the specific total energy is

$$E = u + \frac{1}{2}v^2. \tag{4.8}$$

In these equations ϕ is the gravitational potential, P is the gas pressure, $\boldsymbol{\pi}$ is the viscous stress tensor, $\mathbf{F}$ is the conductive flux, and $\dot{q}$ is any local heating/cooling source. The potential is determined by the Poisson equation

$$\nabla^2 \phi = 4\pi G \rho, \tag{4.9}$$

while in the Navier–Stokes approximation the viscous stress and conductive flux are given by

$$\boldsymbol{\pi} = \rho \nu \left[\nabla \mathbf{v} + (\nabla \mathbf{v})^T - \frac{2}{3}\mathbf{I}\nabla \cdot \mathbf{v}\right] \tag{4.10}$$

$$\mathbf{F} = -\kappa \nabla T. \tag{4.11}$$

Here $\mathbf{I}$ is the identity tensor. The pressure, kinematic viscosity ν, thermal conductivity κ, and temperature T must be determined as functions of ρ and u using constitutive relations to be discussed later.

With the exception of the gravitational terms, the spatial derivatives in Equations (4.5)–(4.7) are gradients or divergences. If we integrate these equations over a spatial volume, these become flux integrals over the surface bounding the volume. Thus the equations can be regarded as local conservation laws. In formulating numerical methods for solving the hydrodynamical equations, it is important to explicitly preserve this conservative property.

Common-envelope interactions, and astrophysical flows more generally, develop shock waves that convert bulk kinetic energy into heat on length scales comparable to the atomic mean free path. From the point of view of the hydrodynamical equations, a shock is essentially a discontinuous jump in density, pressure, and velocity whose magnitudes can be determined by integrating the equations across the shock. These jumps in turn determine the speed of the shock. Numerical simulations inevitably smear shocks over one or more resolution elements, but if they do not locally conserve density, momentum, and energy, the jumps in density, pressure, and velocity across shocks will be incorrect, leading to incorrect shock speeds (Hou & Floch 1994).

4.1.2 Diffusive Transport

Under the conditions present in stellar envelopes (apart from white dwarfs) we can generally neglect the viscous and conductive terms in the hydrodynamical equations. The typical speed of a particle relative to $\mathbf{v}$ is the sound speed c_s, which for an ideal gas with first adiabatic index Γ_1 is given by

$$c_s = \sqrt{\frac{\Gamma_1 P}{\rho}}. \tag{4.12}$$

The molecular viscosity and conductivity have magnitudes

$$\nu \sim c_s \lambda \tag{4.13}$$

$$\kappa \sim c_s \lambda \rho k / m. \tag{4.14}$$

In the envelopes of giant stars, $\rho \sim 10^{-10}$–10^{-5} g cm^{-3} and $T \sim 10^3$–10^6 K. Assuming an ion scattering cross section $\sigma \sim 10^{-20}$ cm^2, the mean free path is $\lambda = m/(\rho\sigma) \sim 10$–$10^6$ cm, while the sound speed $c_s \sim (kT/m)^{1/2} \sim 10^5$–$10^7$ cm s^{-1}. Thus the viscosity is $\nu \sim 10^6$–10^{13} cm^2 s^{-1}, and the conductivity is $\kappa \sim 10^4$–10^{16} cm^4 s^{-3} K^{-1}. Measures of the importance of the dissipative terms in the Navier–Stokes equations relative to the inertial terms are given by the Reynolds and Peclet numbers

$$\mathrm{Re} \equiv \frac{|\nabla \cdot (\rho \mathbf{v}\mathbf{v})|}{|\nabla \cdot \boldsymbol{\pi}|} \sim \frac{vL}{\nu} \tag{4.15}$$

$$\mathrm{Pe} \equiv \frac{|\nabla \cdot [(\rho u + P)\mathbf{v})]|}{|\nabla \cdot \mathbf{F}|} \sim \frac{\rho k v L}{m\kappa}. \tag{4.16}$$

Typical speeds developed during common envelope evolution are transonic to weakly supersonic, so on scales $L \sim R_\odot$ we have Re $\sim 10^3$–10^{12} and Pe $\sim 10^{-2}$–10^{17}. Under most circumstances, then, the inertial terms dominate and we can treat the gas as inviscid and nonconducting. Under these conditions Equations (4.5)–(4.7) are called the Euler equations.

Although this simplifies the fluid equations, it also makes the gas more susceptible to the development of turbulence. In convective stellar envelopes turbulence is created by buoyancy-driven instabilities. These inject energy into the turbulent cascade at scales comparable to the pressure scale height $H_P = P/|\nabla P|$, and given the very large Reynolds number of the flow the turbulent motions are dissipated into heat on scales $\sim \mathrm{Re}^{3/4} \sim 10^9$ times smaller. Numerical simulations typically cannot resolve such a large range of length scales in three dimensions. One approach adopted in turbulence modeling is large eddy simulation (LES), in which the largest scales are directly simulated by solving a low-pass filtered version of the Navier–Stokes equations. The filtering process adds diffusion-like source terms to the equations which encode the behavior of the fluid on scales below the filtering scale. These terms are treated using "subgrid models" that typically are specific to the nature of the type of turbulence (e.g., buoyancy- versus shear-driven). In convective stellar envelopes the effective turbulent viscosity and conductivity can be significantly larger than the values due to atomic interactions, since H_P replaces λ in Equations (4.13) and (4.14). In giant star envelopes $H_P > R_\odot$, so these diffusive coefficients can be $>10^5$ times larger than the atomic values.

Most astrophysical simulations, and in particular most common envelope simulations, do not explicitly include turbulence subgrid models, instead relying on truncation error (Section 4.3.1) to provide the small-scale dissipation needed to turn turbulent motions into heat. While there is some evidence that this "implicit LES" approach is not unreasonable, it does produce turbulent kinetic energy spectra that differ from resolved viscous simulations at high wavenumber, and once flow becomes turbulent the turbulence does not fully decay (Sytine et al. 2000; Aspden et al. 2008). In common envelope problems involving low-mass giant donors, the timescale of the dynamical plunge is typically short enough that convective energy transport is not important, but as the system transitions to a self-regulated phase, the nature of the inspiral begins to depend sensitively on the rate of energy deposition in the envelope (see Chapter 7). Correct treatment of this phase in simulations therefore requires moving beyond implicit LES and incorporating an appropriate subgrid model for convective transport.

4.1.3 Equation of State

If we neglect diffusive terms, the six Equations (4.5)–(4.7) and (4.9) contain seven unknown quantities and so are not closed. Closure is provided by specifying an equation of state (EOS) that gives $P(\rho, u)$. Frequently the ideal gas EOS

$$P = (\Gamma_3 - 1)\rho u = \frac{\rho k T}{\mu}, \tag{4.17}$$

is used for simplicity, where μ is the atomic mass, and $\Gamma_3 = 5/3$ corresponds to zero internal degrees of freedom. This EOS neglects degeneracy, radiation, nonideal effects, dust, and ionization state, and thus it has a limited range of applicability in stellar envelopes. However, for common envelope problems it is a useful first approximation. For the "classical" common envelope problem involving a giant star donor and a compact companion, the companion and/or the donor's core may be degenerate, but the degenerate regions occupy a tiny fraction of the volume and have high densities, so they are often treated by extracting them and replacing them with particles that interact only gravitationally with the rest of the gas. Low-mass giant envelopes usually have small radiation pressure fractions, though radiation pressure can be an important consideration for massive stars. Nonideal effects are most important for planets, brown dwarfs, and very low-mass stars. Dust is most important at low temperatures and high metallicities, primarily through its coupling with radiation. However, ionization state and associated effects like pressure ionization are an important complication even in the classical problem. Unfortunately, including these effects generally precludes simple analytic forms such as Equation (4.17).

The most accurate and comprehensive EOS currently used by many simulation groups is the tabulated EOS provided as part of the MESA stellar evolution code (Paxton et al. 2011, 2013, 2015, 2018, 2019). This EOS blends several approximate equations of state to cover a region in (ρ, T) space spanning $\log(\rho/\mathrm{g\,cm^{-3}}) - 2\log(T/K) + 12 = -10$ to 5.69 and $\log(T/K) = 2.1$–8.2 for metallicities from zero to solar. Additional tables are provided for some non-solar abundance patterns. Interpolation is used for points not included in the tables, and since simulation codes generally require $P(\rho, u)$ rather than $P(\rho, T)$, iteration is often required to determine the pressure or energy.

Since different codes are often used to create the initial donor model and evolve it within a binary system, an important consideration to be aware of is the need for the equations of state used in the various codes to be consistent. If the density and pressure profiles from the initialization code are used to set up the evolution code, the resulting model may be close to hydrostatic equilibrium but will have a different total energy, and its entropy profile will be different, possibly changing its stability characteristics. On the other hand, to preserve the stability characteristics of the original model, one can integrate the hydrostatic equilibrium equations together with the temperature gradient equation, requiring that $\nabla - \nabla_{\mathrm{ad}}$ with the new EOS match the profile from the stellar evolution code and computing $\rho(P, T)$ using the new EOS. However, the resulting initial model will have a different total mass and radius. Details of this approach are discussed by Ohlmann et al. (2017).

4.1.4 Gravity

In global common envelope simulations the potential ϕ includes contributions from the donor star's core and envelope and the companion, requiring us to numerically solve Equation (4.9). The geometry of the binary orbit and the ejected envelope prevents us from reducing the problem to 1D spherical symmetry or 2D axisymmetry, so a 3D Poisson solver must be used. In addition, the appropriate boundary condition on the potential ($\phi \to 0$ at infinite separation) requires special treatment for algorithms that are based on the differential form of the Poisson equation (such as Fourier transform or multigrid techniques).

As noted above, the donor's core and the companion generally have densities much greater than that of the envelope and occupy a small fraction of the volume, even if they are not degenerate. For example, a 5 $M_\odot$ star at the tip of the red giant branch can have a stellar radius of 160 $R_\odot$ and a helium core radius of only 0.1 $R_\odot$.

Thus it is generally not feasible to resolve them spatially in a simulation that also tracks the binary orbit and the envelope ejection. Moreover, the common envelope interaction leaves the hydrostatic equilibrium of the core regions largely untouched (though depending on the thermal timescale they might be expected to expand or contract as the envelope is ejected). Solving the hydrodynamical equations in the cores would thus be an error-prone exercise in maintaining hydrostatic equilibrium over many dynamical times. For these reasons most common envelope simulations replace the core of the donor (and often the companion as well) with a particle that interacts only gravitationally with the envelope gas. The motion of these particles is integrated numerically alongside the hydrodynamics of the envelope. The core particle is not a point mass but corresponds to a potential with a finite softening width; the exact form corresponds to the particle's assumed density profile. It is important to emphasize that the numerical core is not the same as the physical core of the star (whether defined via hydrogen mass fraction or any other means), but rather is chosen to be smaller than the expected final binary separation but large enough to make the simulation tractable. Of course, if the binary interaction results in merger, this core approximation must be abandoned (at least for the later stages of the interaction). Numerical tests by Iaconi et al. (2017) and Chamandy et al. (2018) suggest that the numerical core radius should be smaller than 1/5 of the binary separation, and perhaps smaller still, to achieve convergence for systems that do not merge. Nonconvergence produces final separations that are too large and unbound masses that are too small. We discuss convergence further in Section 4.3.2.

Ohlmann et al. (2017) describe a technique for self-consistently introducing core particles into the donor model that is now used by most common envelope simulation groups in some form. A stellar evolution code is used to produce 1D radial profiles of density, pressure, and composition, and a numerical core radius $r_{\rm core}$ is chosen. The functional form of the density profile $\rho_{\rm core}(r)$ associated with the core particle is specified; its corresponding gravitational acceleration profile is

$$g_{\rm core}(r) = -\frac{Gm_{\rm core}(r)}{r^2} = -\frac{4\pi G}{r^2}\int_0^r dr'\, r'^2 \rho_{\rm core}(r'). \tag{4.18}$$

Because the core density profile does not match the stellar evolution model, some gas remains at radii $r < r_{\rm core}$. This gas is treated as a polytrope with density $\rho_{\rm poly}(r)$ and pressure $P_{\rm poly}(r) = K\rho_{\rm poly}(r)^{1+1/n}$ satisfying the hydrostatic equilibrium condition

$$\frac{1}{4\pi G r^2}\frac{d}{dr}\left[r^2\left(\frac{1}{\rho_{\rm poly}(r)}\frac{dP_{\rm poly}}{dr} - g_{\rm core}(r)\right)\right] + \rho_{\rm poly}(r) = 0 \tag{4.19}$$

for a given value of n. Using appropriate scalings this can be cast as a modified version of the Lane–Emden equation. The constant K and central density $\rho_{\rm poly}(0)$ are determined by requiring that $\rho(r)$ and its derivative be continuous with the stellar evolution model at $r = r_{\rm core}$. As with the ordinary Lane–Emden equation, the convective stability of the polytrope is determined by n and the adiabatic index $\gamma_{\rm ad}$; convective stability holds if $n > \frac{1}{\gamma_{\rm ad} - 1}$. The mass $M_{\rm core} \equiv m_{\rm core}(r_{\rm core})$ of the core particle can be specified or left as a free parameter to be constrained by requiring that

$$m(r_{\rm core}) = M_{\rm core} + m_{\rm poly}(r_{\rm core}), \tag{4.20}$$

where $m(r)$ and $m_{\rm poly}(r)$ are the enclosed mass profiles of the original model and polytrope, respectively. If $M_{\rm core}$ is independently specified, in general one must re-integrate the hydrostatic equilibrium of the model outward from $r_{\rm core}$ to determine the pressure and other thermodynamic quantities, possibly constraining $\nabla - \nabla_{\rm ad}$ to preserve convective properties as described in the previous section.

Once initialized, during a simulation the core particles' gravitational acceleration is directly added to the acceleration determined by solving the Poisson equation for the remaining gas to determine the total gravitational acceleration of the gas. The equations of motion for the core particles include their joint gravitational interaction and the gravitational acceleration due to the gas. The particle core radius $r_{\rm core}$ and the effective softening length h of the gas potential (either explicitly imposed as in SPH or implicitly related to the zone spacing as in AMR) should be chosen so that $r_{\rm core} \gtrsim 2 - 3h$ to avoid unphysical kicking of the core particles due to noise in the gas density field. This is of particular concern for the particle representing the donor's core, since (at least initially) it is close to the minimum of the gas potential field.

4.1.5 Accretion

During the plunge and slow inspiral phases of common envelope evolution, a compact companion star accretes matter from the envelope. Because the size of the companion and its accretion disk are generally much smaller than a simulation's resolution elements, measurement of the accretion rate can only be done approximately. This has implications not only for estimates of the gravitational drag force on the companion but also for modeling of energetic feedback (Section 3.3.1).

The Bondi–Hoyle–Lyttleton radius describes the length scale on which the gravitational field of an accretor of mass $M_{\rm comp}$ affects the surrounding gas:

$$\begin{aligned} R_{\rm BHL} &= \frac{2GM_{\rm comp}}{v_{\rm orb}^2(1 + \mathcal{M}^{-2})} \\ &\approx 3850\left(\frac{M_{\rm comp}}{M_\odot}\right)\left(\frac{v_{\rm orb}}{10\ {\rm km\ s^{-1}}}\right)^{-2}(1 + \mathcal{M}^{-2})^{-1}\, R_\odot. \end{aligned} \tag{4.21}$$

Here $v_{\rm orb}$ is the orbital velocity of the companion and $\mathcal{M}$ is the upstream Mach number in the companion's rest frame. This is much larger than typical resolution elements in CE simulations ($<R_\odot$) and even some donor stars, so the region from which the companion could accrete given sufficient time is easily resolved. However, for compact companions the size of the accretor can be smaller than 1% of the resolution length scale, so the accretion flow cannot be directly simulated.

To handle this situation, a spherical "control surface," or alternatively an embedded inner boundary, of radius $R_{\rm cs}$ is introduced around the companion in order to treat the accretion flow (Ricker & Taam 2008). The rates of mass and momentum accretion (used to measure drag) through this surface are measured at any instant using the surface integrals

$$\dot{M} = -R_{\mathrm{cs}}^2 \int_{\mathrm{cs}} d\Omega\, \rho \mathbf{v} \cdot \hat{\mathbf{n}} \tag{4.22}$$

$$\dot{\mathbf{P}} = -R_{\mathrm{cs}}^2 \int_{\mathrm{cs}} d\Omega\, \rho \mathbf{v}(\mathbf{v} \cdot \hat{\mathbf{n}}), \tag{4.23}$$

where velocities are measured in the companion's rest frame. The details of the integration vary from code to code, but generally R_{cs} is taken to be at least a few resolution elements across, to reduce noise in the estimate, but not much more, so as not to capture gas that will not accrete during a time step. A control surface is a passive construct allowing for measurement without altering the solution. Without modification the flow develops a back pressure that resists further accretion, even at sub-Eddington rates, making the measured fluxes a lower limit if we neglect other physics (such as neutrino cooling) that might come into play. Sink particle treatments based on Krumholz et al. (2004) actively remove material from the simulation, adding it to the companion's mass, after each time step to prevent this from occurring (MacLeod & Ramirez-Ruiz 2015; Chamandy et al. 2018). In neither case can the flow inside R_{cs} be considered numerically reliable, but the aim is to reproduce the flow around the accretor as accurately as possible and to measure the accretion rate with an error that is uncontrolled, but perhaps optimistically no worse than a factor of two.

These uncertainties are amplified when we allow for the imposition of jet or wind feedback (e.g., Chamandy et al. 2018; Shiber 2018; López-Cámara et al. 2019). In this case, either a fixed amount or some adjustable fraction (typically $\mathcal{O}(10\%)$) of the mass-energy accreted during a time step is returned to the simulation as a local source term or (less commonly) as an imposed inner boundary flux. For local source terms, the feedback mass and energy are added to a cylindrical or conical region just outside the control surface. The fixed parameterization of the feedback characteristics is almost certainly incorrect but allows for exploration of the effect of different amounts of feedback. Moreover, the positioning of local source terms next to the control surface affects the measured accretion rate. Simulations including accretion and feedback should therefore be considered exploratory at this time and not relied upon for detailed theory predictions.

4.1.6 Radiation

Radiation transport can have a number of effects on common envelope evolution. Most uniquely, it removes energy from the envelope. Through its influence on the equation of state it can affect the dynamical stability of the envelope. It also serves as a source of pressure. These effects are not significant for all common envelope systems, but for some donors, such as massive stars, they can make a big difference. Outside of the equation of state, radiation transport effects have only begun to be considered in common envelope simulations (Ricker et al. 2019).

In the geometrical optics limit, the propagation of photons and their interaction with matter is described by the radiation transfer equation (Mihalas & Weibel Mihalas 1984). The solution to the monochromatic (single-frequency) form of this equation is the radiative intensity $I_\nu(\mathbf{x}, \hat{\mathbf{n}}, t)$, which is a function of frequency ν, position $\mathbf{x}$, direction $\hat{\mathbf{n}}$, and time t. Because of the number of independent variables and the complexity of the emission and absorption coefficients as functions of gas properties, the numerical solution of the radiation transfer equation in full generality is usually not feasible. We therefore look for simplifying approximations. Deep inside the stellar photosphere, the most important approximations include local thermodynamic equilibrium (LTE: the radiation has a blackbody spectrum, and its local temperature is the same as that of the gas) and collisionality (the photon mean free path is short compared to the pressure scale height). These assumptions break down outside the photosphere. We also assume nonrelativistic velocities and weak gravitational fields. These assumptions are valid throughout the bulk of a non-degenerate donor star's envelope but clearly break down in the immediate vicinity of a neutron star or black hole companion.

Under collisional LTE conditions we can derive a widely used (and reviled) approximation known as the diffusion approximation. As with the Navier–Stokes equations, we can take velocity moments of the radiation transfer equation to derive a set of moment equations that require a closure. Since the photon velocity is $c\hat{\mathbf{n}}$, the moments of interest are (upon integrating over frequency so that, e.g., $I = \int_0^\infty d\nu\, I_\nu$)

$$\text{energy density } u_{\text{rad}} = \frac{1}{c} \int d\Omega\, I \tag{4.24}$$

$$\text{radiative flux } \mathbf{F}_{\text{rad}} = \int d\Omega\, \hat{\mathbf{n}} I \tag{4.25}$$

$$\text{radiation pressure } P_{\text{rad}} = \frac{1}{3c} \text{Tr} \int d\Omega\, \hat{\mathbf{n}}\hat{\mathbf{n}} I \tag{4.26}$$

where $d\Omega$ is the differential element of solid angle in the direction of $\hat{\mathbf{n}}$ and Tr is the trace. To obtain the diffusion approximation, we expand I in inverse powers of the optical depth (Mihalas & Weibel Mihalas 1984). At zeroth order, the intensity is isotropic and equal to the blackbody intensity at temperature T, so $u_{\text{rad}} = aT^4 = 3P_{\text{rad}}$ and $\mathbf{F}_{\text{rad}} = 0$. At this order the radiation energy density and pressure can be added to those of the gas, and under nonrelativistic conditions we can treat the combined fluid using the Navier–Stokes equations. At this order the only ways energy can be transported are via advection (fluid motion), viscous heating by the gas, or thermal conduction (in degenerate material).

At the next order of approximation, the radiation intensity has a nonzero first angular moment. The radiative flux is then given by

$$\mathbf{F}_{\text{rad}} = -\frac{c}{\rho\bar{\kappa}} \nabla P_{\text{rad}}. \tag{4.27}$$

The radiative flux is added to the conductive flux in the combined Navier–Stokes energy equation. Here the Rosseland mean opacity is defined as

$$\bar{\kappa} \equiv \frac{\int d\nu \frac{dB_\nu}{dT}}{\int d\nu \frac{1}{\kappa_\nu}\frac{dB_\nu}{dT}}, \tag{4.28}$$

where $B_\nu(T)$ is the Planck function and κ_ν is the frequency-dependent opacity of the gas. In numerical calculations this opacity is determined as a function of density, temperature, and composition using interpolation from tables such as the OPAL tables (Iglesias & Rogers 1996). The form of the Rosseland mean reflects the fact that, in optically thick material, the continuum spectrum dominates radiative transport.

The diffusion approximation is simple and forms the basis of the radiative temperature gradient equation in the stellar structure equations. However, in a dynamical, multidimensional context this approximation has some major drawbacks. Most importantly, it applies only under optically thick conditions. In the vicinity of the photosphere and beyond it overpredicts the radiative flux, partly because the intensity becomes highly anisotropic there and partly because the effective mean opacity in optically thin regions is better represented by the Planck mean. A common fix-up is to artificially limit the flux in regions of strong temperature gradients so that the effective information propagation speed does not exceed the speed of light (Levermore & Pomraning 1981). For example, the Levermore–Pomraning flux limiter takes

$$\mathbf{F}_{\rm rad} = -cu_{\rm rad}\lambda(|\mathbf{R}|)\mathbf{R}, \tag{4.29}$$

where

$$\mathbf{R} \equiv \frac{\nabla u_{\rm rad}}{\rho\bar{\kappa}u_{\rm rad}} \tag{4.30}$$

and

$$\lambda(R) = \frac{1}{R}\left(\coth R - \frac{1}{R}\right). \tag{4.31}$$

In the limit of small radiative flux (optically thick), $R \to 0$ and $\lambda(R) \to \frac{1}{3}$, recovering Equation (4.27). In the limit of a large flux (at the photosphere), $R \to \infty$ and $\lambda(R) \approx 1/R$, giving $\mathbf{F}_{\rm rad} = -cu_{\rm rad}\hat{R}$.

However, flux-limited diffusion does not treat the flux correctly in optically thin regions outside the photosphere where the temperature gradient may be small or the intensity field is far from isotropic. Worse, in the presence of hot artificial "fluff" material (to be discussed later) it may yield an energy flux into the star.

One desirable output from radiation transfer would be the light curve and spectral evolution of a common envelope system. However, in practice these are challenging to compute. As emphasized by Galaviz et al. (2017), in most common envelope simulations the photosphere region is not resolved, so the correct surface flux (and thus the effective temperature) cannot be reliably predicted. Resolution elements that have a characteristic size of Δr near the surface have very large optical depths $\tau \equiv \rho\kappa\Delta r$. Their properties therefore do not reflect those that would exist at optical depths $\tau \approx 1$, i.e., at the locations from which photons would be expected to escape. One promising approach is to fit atmosphere models to the elements near the surface to try to estimate the effective temperature.

4.1.7 Other Sources of Complication

Other physical processes, which may or may not play a role in specific common envelope events, include generation of magnetic fields, explosive nuclear reactions, and dust formation. These are potentially relevant processes, but as these areas are relatively unexplored we will not go into the many numerical details here.

4.2 Numerical Methods

Because of the large range of timescales involved in stellar evolution, it is necessary to use multiple simulation codes to study common envelope evolution. The initial model for the donor star must be generated as a one-dimensional model using a 1D stellar evolution code, because the evolutionary timescale preceding the initiation of mass transfer lasts many dynamical times. Because of the very aspherical structure of the envelope during the CE phase, this model must be transferred in some way to a 3D hydrodynamical code to follow the plunge-in phase.

As in other areas of astrophysics, both mesh- and particle-based simulation methods have been applied for the hydrodynamical simulations. Because of the envelope's compressibility, these methods generally have employed some form of shock capturing together with explicit time integration. For convenience, Table 4.1 shows some of the simulation codes in current use for 3D hydrodynamical

Table 4.1. Numerical Codes in Use as of 2020 for 3D Simulations of Common Envelope Evolution

Code	Hydrodynamics	Adaptivity	First Use	Method Paper
FLASH	PPM	AMR	Ricker & Taam (2008)	Fryxell et al. (2000)
Enzo	PPM	AMR	Passy et al. (2012)	Bryan et al. (2014)
AstroBEAR	PPM	AMR	Chamandy et al. (2018)	Cunningham et al. (2009)
Athena++	PPM	AMR	MacLeod et al. (2018)	Stone et al. (2020)
GADGET	SPH	&	Pakmor et al. (2012)	Springel (2005)
SNSPH	SPH	&	Passy et al. (2012)	Fryer et al. (2006)
StarSmasher	SPH	VMP	Lombardi et al. (2011)	Gaburov et al. (2010)
Phantom	SPH	&	Reichardt et al. (2019)	Price et al. (2018)
AREPO	PLM	MM	Ohlmann et al. (2016)	Springel (2010)
MANGA	PLM	MM	Prust & Chang (2019)	Chang et al. (2017)

Notes. Acronyms for numerical methods are defined in the text. "First use" refers to the first paper in which the code was applied to common-envelope simulations. "Method paper" refers to the paper in which numerical methods and tests were first presented. VMP stands for varying mass particles.

simulations of common envelope evolution. As a visual aid for the discussion to follow, in Figure 4.1 we illustrate the different ways in which adaptive mesh refinement (AMR), moving mesh (MM), and smoothed particle hydrodynamics (SPH) codes represent the same distribution of matter.

Finally, as the system enters a self-regulated inspiral, a different technique must be adopted (e.g., a 1D drag model) to cope with the lengthened evolutionary timescale. A number of codes and techniques have been employed for each of these phases. Here we review some of those which have been most commonly used for the first two phases; for a discussion of the self-regulated phase see Chapter 7.

4.2.1 Generating Initial Conditions

Stellar Evolution Models

The earliest common-envelope simulations used initial donor models drawn from 1D single-star evolutionary calculations made, for example, using variants of Peter Eggleton's (1971) code. Although mass transfer episodes may occur prior to the common envelope interaction under study, these simulations neglected the effect of binary interactions in evolving toward the initial model for the common envelope phase. More recent simulations have predominantly used models constructed using the MESA stellar evolution code (Paxton et al. 2011, 2013, 2015, 2018, 2019). These too have neglected pre-common envelope binary interactions, i.e., have used unperturbed single-star models as input. Stellar evolution codes, including MESA, are able to follow the mass-transfer phase until the initiation of dynamical instability. In principle it would be far preferable not to use unperturbed single-star models as input.

In addition, stellar evolution models used to initialize common-envelope simulations generally have neglected rotation. For slow rotation (less than ~10% of the critical rotation rate) the 1D stellar evolution codes can include axisymmetry via the shellular approximation (Meynet & Maeder 1997; Maeder & Meynet 2000), but this

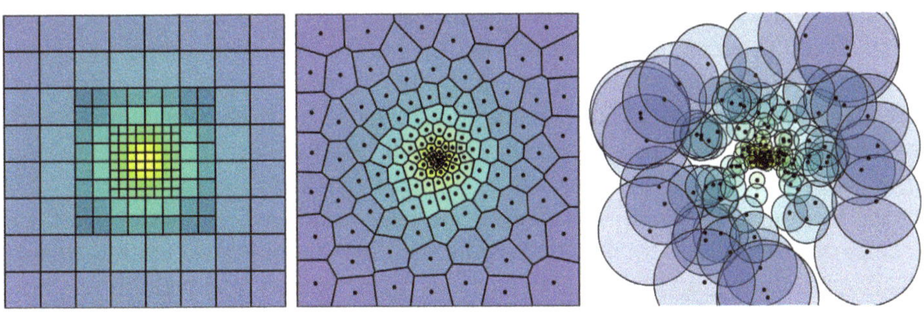

Figure 4.1. Different methods used for representing gas quantities for multidimensional hydrodynamics simulations of common-envelope evolution. The same centrally concentrated density profile is used for each frame, and the central resolution in each case is comparable. Left: nested-grid or adaptive mesh refinement (AMR) with three mesh levels. Middle: moving mesh (MM) using a regularized Voronoi tessellation. Right: smoothed particle hydrodynamics (SPH). For clarity, SPH is shown with a much smaller number of neighbors per particle than is customarily used.

will not be appropriate for rapidly rotating donors, such as stars undergoing Roche-lobe overflow (RLOF). In addition to the geometry of the donor, rotation can have effects on its density, temperature, and chemical profiles. The effects of these changes on common-envelope evolution are largely unexplored.

Mapping into Three Dimensions

Stellar evolution codes produce 1D radial profiles of density, pressure, and composition that represent a hydrostatic stellar model on the brink of unstable RLOF when paired with an appropriate companion at a given separation. To follow the common-envelope interaction they must be mapped into three dimensions in a way that preserves this condition as much as possible. This is an inherently difficult problem: to demonstrate that the initial conditions are in hydrostatic equilibrium, they must be maintained by a simulation code for multiple dynamical times. However, a system on the brink of common envelope mass transfer is by definition dynamically unstable.

Often simulators have finessed this problem by mapping the 1D model into 3D, relaxing it in isolation, and then adding the binary companion (e.g., Sandquist et al. 1998). However, the resulting systems have unrealistic total energies. A better approach is to map the 1D model into 3D while the binary orbit is still some ways from RLOF. The processes that tip the system toward unstable mass transfer can require many orbits to have their effect. In order to bring the model up to the brink of RLOF, two techniques have been used. In one, the donor is relaxed at a sequence of decreasing orbital separations until the final separation is achieved (Lombardi et al. 2011). In the other, an artificial drag force is applied to bring the stars together and speed up the start of RLOF (Rasio & Livio 1996). These techniques offer the possibility of a more realistic approach to unstable mass transfer.

However the approach to the common-envelope phase is treated, the donor model must be mapped into the 3D simulation code in a manner that preserves the total mass, momentum, and energy of the star. For a uniform-grid or adaptive mesh refinement (AMR) code, the simplest approach is to linearly interpolate to each grid point from the 1D stellar model, rescaling densities, velocities, and energies at the end to match the total conserved quantities in the 1D model. However, this approach can produce unphysical step features in the 3D distributions, particularly at the center of the star and in regions of strong gradients or where mesh refinement level changes (Chen et al. 2011). For smoothed particle hydrodynamics (SPH) codes, mapping the 1D model profiles is more difficult because they specify the values of kernel sums taken over an unknown set of SPH particle values at unknown SPH particle locations. Different techniques have been employed, including placing the particles on a hexagonal mesh and allowing their masses and/or occupancy fractions to vary and building up the star with particles distributed in spherical shells either as a glass or using HEALPix tiling (Diehl et al. 2015; Joyce et al. 2019). The hex-grid approach reduces the development of SPH pairing instabilities but is noisy, while the spherical shell technique reduces noise but is not the lowest-energy configuration. In both cases the model must be rescaled to match the 1D model total quantities, and if SPH particle masses are allowed to vary an appropriate kernel density sum must be employed (Rosswog 2009).

Initial model mapping usually includes a period of forced velocity damping before the binary is allowed to enter common envelope. For example, Ohlmann et al. (2017) apply a damping acceleration

$$\dot{\mathbf{v}}_{\mathrm{damp}} = -\mathbf{v}/\tau, \tag{4.32}$$

where the damping timescale is given by

$$\tau(t) = \begin{cases} \tau_1 & t < 2t_{\mathrm{dyn}} \\ \tau_1\left(\dfrac{\tau_2}{\tau_1}\right)^{(t-2t_{\mathrm{dyn}})/3t_{\mathrm{dyn}}} & 2t_{\mathrm{dyn}} < t < 5t_{\mathrm{dyn}}, \\ \infty & t > 5t_{\mathrm{dyn}} \end{cases} \tag{4.33}$$

where τ_1 is 10% and τ_2 is 100% of the dynamical timescale t_{dyn}. The aim is to gradually reduce the damping to zero, leaving a nearly hydrostatic star. However, donors entering common envelope generally have convective envelopes. Thus while they may be in hydrostatic equilibrium when averaged over many dynamical times, they are unstable to the development of small-scale convective motions that produce energy transfer (mediated implicitly by numerical dissipation as described earlier). The Reynolds number of this turbulence, and thus the resolution required to directly simulate it, is typically very high. However, to the extent that we can trust implicit LES, and when the thermal adjustment timescale is long compared to the dynamical timescale, the convective velocities that naturally arise when damping is turned off (and artificial viscosity is not used) are highly subsonic and do not affect the outcome of a common envelope simulation. Convection is much more of a problem when the envelope is close to the Eddington limit and/or the thermal timescale is short, as in massive star envelopes (Ricker et al. 2019). In these cases radiation transport must be included, even at the relaxation phase, so that the convective velocities are not strongly overestimated.

4.2.2 Grid-based Methods

The earliest multidimensional common envelope simulations used classical numerical methods with artificial viscosity on uniform or moving meshes. For example, the 2D axisymmetric simulations described by Taam & Bodenheimer (1989) used a code based on the first-order donor-cell method (Black & Bodenheimer 1975). Later 2D grid-based work (Yorke et al. 1995) used a staggered-mesh code with some second-order features (Rozyczka 1985). The earliest 3D grid-based simulations used a "pseudoparticle" method similar to SPH (Livio & Soker 1988), while later 3D work by Sandquist et al. (1998) used a nested-grid code with second-order space and first-order time accuracy (Burkert & Bodenheimer 1993).

This early work showed that envelope ejection is highly non-axisymmetric and somewhat concentrated in the orbital plane. It also demonstrated the importance of large-scale tides in driving orbital dissipation during the initial plunge-in, after which the inspiral slows and becomes more driven by the local tidal field near the companion. This slower evolution is not efficiently followed by current 3D hydrodynamical simulations (Chapter 7).

Even low-resolution and low-order techniques sufficed to establish envelope asymmetry in multidimensions. However, to probe the transition from the plunge-in phase to the self-regulated inspiral phase, and in particular to treat cases in which the donor is a supergiant or the mass ratio is close to unity, required more modern shock-capturing techniques with some form of adaptive resolution control. In general these needs require simulations to take more time steps and use higher resolution, so an improvement beyond first order truncation error and the ability to efficiently deploy higher resolution were needed to make further progress (see Section 4.3 for more discussion on resolution and error).

Numerical methods for compressible hydrodynamics underwent something of a revolution around 1980, though because of the complexity of the new methods some time elapsed before they were widely adopted in astrophysics. In particular, a new class of high-order methods based on extensions of Godunov's (1959) first-order method were developed to dramatically reduce numerical dissipation and capture shocks on meshes using as few as 1–2 zones. As "finite volume" schemes these new methods also explicitly conserved mass, momentum, and energy to machine precision, guaranteeing correct shock speeds. In a finite volume method, the quantities evolved by the code are zone averages rather than point values of density, pressure, velocity, etc. For example, the continuity equation may be written exactly for a 1D uniform Cartesian mesh with spacing Δx using

$$\frac{d\bar{\rho}_i}{dt} = -\frac{\overline{\rho v}_{i+1/2} - \overline{\rho v}_{i-1/2}}{\Delta x}, \tag{4.34}$$

where $\bar{\rho}_i$ is the average density in zone i and $\overline{\rho v}_{i+1/2}$ is the mass flux averaged over the interface between zones i and $i+1$. Finite volume methods differ in how they perform the time integration for the left-hand side and the interpolant reconstruction of the fluxes on the right-hand side. Godunov's method assumes the fluid quantities are piecewise constant and uses the solution to the Riemann problem at each zone interface to determine time-advanced fluxes.

The MUSCL approach (Monotonic Upstream-centered Scheme for Conservation Laws; van Leer 1979) provided a template for extending Godunov's scheme to higher order using linear or quadratic polynomial reconstruction of the zone-edge values used to calculate hydrodynamical fluxes. MUSCL schemes apply various forms of fitting coefficient limiting to ensure the numerical solution is "total variation diminishing" (TVD), a condition that is required to avoid dispersive errors at flow discontinuities. In addition, they modify the input states to account for higher-order variation in the Riemann problem solution. Although MUSCL is most often associated with the second-order "piecewise linear method" (PLM), the widely used "piecewise parabolic method" or PPM (Colella & Woodward 1984) is also a MUSCL scheme. By using third-order calculations in some areas, PPM achieves lower dissipation than the purely linear variant, though for performance reasons it keeps some second-order aspects and thus exhibits second-order truncation error.

When discussing Godunov-type simulation codes, an important consideration that is sometimes overlooked is the fact that when a nonideal EOS is used the Riemann solver must account for varying and unequal adiabatic exponents, such as are encountered in regions of partial ionization. Since equation of state subroutines typically use Newton iteration to obtain thermodynamic quantities, iterating the EOS to convergence within each step of a Newton loop used to solve the Riemann problem can be expensive. A commonly used approximation described by Colella & Glaz (1985) is to treat the adiabatic indices as advected quantities whose transport equations are included as part of a solution of the Riemann problem. However, if the EOS is not convex or the adiabatic indices vary significantly this approach can be inaccurate. To circumvent these problems, Chen et al. (2019) have recently constructed a Riemann solver for general equations of state based on the HLLC approximation to the Riemann problem described by Toro et al. (1994). This is a promising direction, though the solver is still considerably slower than the perfect-gas HLLC method.

Adaptive Mesh Refinement

A second development in the 1980s involved the application of multi-resolution mesh techniques to accelerate the solution of partial differential equations on grids. Originally these ideas were applied to the solution of elliptic equations (such as the Poisson equation) using multigrid (Brandt 1977), but soon they were applied to create adaptive mesh refinement (AMR) methods for parabolic (diffusive, in our context) and hyperbolic (advective) equations (Berger & Oliger 1984; Berger & Colella 1989). In AMR, the computational domain is divided into mesh patches, some of which have higher resolution and overlay coarser patches in "interesting" parts of the domain. These patches are created and destroyed over the course of a simulation in response to the changing solution, e.g., to provide high resolution of a traveling shock only in its vicinity. Quirk (1991) developed a simplified AMR algorithm called "oct-tree" refinement that is popular because it can be made highly scalable on parallel computers. However, in general parallel AMR codes are difficult to write and debug, so astrophysical AMR codes did not begin to appear until the 1990s (Bryan 1996).

Ricker & Taam (2008, 2012) first applied PPM and AMR to the common-envelope problem with 3D simulations of low-mass systems using the FLASH code (Fryxell et al. 2000), while around the same time Passy et al. (2012) applied the 3D AMR code Enzo (O'Shea et al. 2004; Bryan et al. 2014) to the problem using uniform meshes. More recently the AstroBEAR AMR code (Cunningham et al. 2009) has been applied to common-envelope simulations, with a focus on accretion onto a compact companion and the effect of its feedback on the envelope (Chamandy et al. 2018). Athena++ (Stone et al. 2020) has also been used by MacLeod et al. (2018) to study the onset of common-envelope events, though without including the self-gravity of the envelope.

Although AMR techniques allow simulations to put high-resolution meshes only where they are needed, the creation and destruction of mesh patches introduces interpolation errors and a dependence on the criteria used to choose whether to refine or derefine a patch. In some cases the refinement criterion can produce artifacts, especially if refinement is turned off in some regions (as it often is in the low-density "fluff" described below). Moreover, although fluxes are adjusted at interfaces between fine and coarse patches to preserve explicit conservation, the discontinuous jump in mesh spacing at such locations reduces the local order of truncation error by one. Thus a shock which passes through a refinement boundary will produce a reflection. A final criticism of AMR methods is that they are generally Eulerian, i.e., the mesh reference frame is fixed and the fluid moves with respect to it. Because of this, truncation error produces artificial mixing of fluid components with diffusive coefficients that depend on the velocity of the fluid with respect to the mesh. Since this mixing is greatest at sharply refined contact discontinuities, a lack of resolution at these locations can break Galilean invariance (Robertson et al. 2010). We further discuss truncation error and mixing in Section 4.3.1.

Unstructured (Moving) Mesh Methods

An alternative (Lagrangian) approach is for the mesh to move with the fluid. (It is also possible to have meshes that are Lagrangian in some regions and Eulerian in others.) This has the benefit of significantly reducing artificial mixing, though in order for shocks to exist there still must be energy dissipation, either due to truncation error or explicit viscosity terms It also has the benefit of allowing high resolution in regions where the density is highest, as do the SPH methods discussed next. Many 1D stellar evolution codes in fact can use Lagrangian meshes, since stars dramatically expand and contract over their lifetimes. However, in multidimensions a structured Lagrangian mesh (that is, one that is amenable to i, j, k indexing) following a turbulent flow will twist and stretch to follow the flow, producing zones whose sides have dramatically different lengths. Since the truncation error in each coordinate direction depends on the zone spacing, this means that the error becomes highly anisotropic in such regions.

These difficulties can be mitigated by employing an unstructured mesh, in which mesh zones have a fixed number of neighbors but cannot be indexed. Unstructured meshes have long been popular in engineering because they can be used to follow irregular or curved boundaries, but they only become widely used in astrophysics during the past decade. The AREPO code (Springel 2010) is a prominent example that was originally developed for cosmological simulations but has been applied to common envelope simulations by Ohlmann et al. (2016, 2017). AREPO uses Voronoi tessellation to determine the placement of mesh elements given a collection of moving point particles that represent the fluid, an approach that allows for efficient mesh generation and adaptation. In regions where the flow would cause a structured Lagrangian mesh to twist and stretch, a simple mesh regularization algorithm is used to keep mesh elements compact. Mesh elements can also be subdivided in regions where additional resolution is desired, independently of the gas density. While these capabilities introduce interpolation errors that make the code not truly Lagrangian, the resulting moving-mesh (MM) algorithm is powerful, efficient, and flexible. In addition to the AREPO group, Prust & Chang (2019) have recently used an independently developed MM code called MANGA (Chang et al. 2017) to perform common envelope simulations including recombination energy.

"Fluff"

Whether Eulerian or Lagrangian, grid-based codes generally apply the same numerical update scheme in all parts of the computational volume, including "vacuum" regions that do not contain gas. Such regions introduce difficulties because hydrodynamical schemes require nonzero densities in order to compute the sound speed. To handle vacuum regions, simulators initialize them with a low-density medium sometimes referred to as the "fluff."

Fluff must have certain properties to avoid adversely affecting the simulation. Its density must be low enough that the total mass in fluff regions is small compared to the masses of the stars. Its pressure must be close to, but less than, the surface pressures of the stars to avoid generating strong shocks in the fluff and to avoid altering the hydrostatic equilibrium of the stars through confinement. Since satisfying the first criterion usually means that the fluff density is much smaller than the surface densities of the stars, together these criteria mean that the temperature of the fluff will be high. This creates a new problem: if the fluff temperature is too high, it will make the integration time step required for stability (Section 4.3.1) too short. A hot fluff also causes problems due to its opacity when calculating light curves (Galaviz et al. 2017) or doing flux-limited radiation diffusion (Ricker et al. 2019).

Choosing appropriate fluff properties requires balancing these constraints and performing experiments with different values to test the sensitivity of results to the fluff. Finding a way to eliminate the need for fluff, for example by introducing a free internal boundary, would be a useful improvement to existing codes, though fluff does not appear to cause insurmountable problems if treated carefully.

Including Self-gravity in Grid Codes

Including the gravitational acceleration due to the gas and the numerical core particles requires that we solve the Poisson Equation (4.9) at each time step. The field due to the gas is generally computed by solving this equation for the potential $\phi_{\rm gas}$ on the grid and then numerically differencing it to find the acceleration $\mathbf{g}_{\rm gas} = -\nabla\phi_{\rm gas}$. The acceleration due to the cores is added directly using their positions at a given time[1] to find the total acceleration at a given point $\mathbf{x}$:

$$\mathbf{g}(\mathbf{x}) = \mathbf{g}_{\rm gas}(\mathbf{x}) + g_{\rm core}(|\mathbf{x} - \mathbf{x}_{\rm core}|)\frac{\mathbf{x} - \mathbf{x}_{\rm core}}{|\mathbf{x} - \mathbf{x}_{\rm core}|} + g_{\rm comp}(|\mathbf{x} - \mathbf{x}_{\rm comp}|)\frac{\mathbf{x} - \mathbf{x}_{\rm comp}}{|\mathbf{x} - \mathbf{x}_{\rm comp}|}. \quad (4.35)$$

Here the acceleration due to the donor core particle, $g_{\rm core}$, is given by Equation (4.18) for a core particle located at $\mathbf{x}_{\rm core}$. The corresponding function for the companion particle (subscript "comp") is usually taken to have the same form. For converged

[1] Ricker & Taam (2008, 2012) used a different approach in which they treated the cores as rigid spherical clouds of particles whose masses were interpolated onto the grid. The combined grid-mapped particle density and gas density were then used as the source in the Poisson equation.

results it is usually necessary to ensure that characteristic width of the density function associated with each particle is resolved by the grid with at least three zones.

Most AMR codes use some form of multigrid algorithm to solve the Poisson equation on the mesh (see, e.g., Trottenberg et al. 2001). These methods use the coarser mesh levels to compute long-range interactions and finer mesh levels to compute short-range interactions. Within each level, uniform-grid techniques such as Gauss–Seidel relaxation or Fast Fourier Transforms are used, with boundary conditions imposed by the next coarser level. The results are iteratively interpolated to the next finer level or restricted to the next coarser level as necessary to achieve an iteratively converged result.

Since the binary is treated as an isolated system, the appropriate external boundary condition is not Dirichlet (in which the potential goes to zero on the boundary), but rather must be consistent with $\phi \to 0$ at infinity, with zero density outside the computational volume. Different techniques to accomplish this exist, but the most memory-efficient involves two Poisson solves, one for the interior source distribution with Dirichlet boundary conditions, and one for an image potential correction (James 1977). Isolated boundary conditions for the potential are, of course, inconsistent with the outflow boundary conditions generally used for the hydrodynamics, as material lost through the external boundary should not immediately stop exerting a gravitational force on the material still inside the computational volume. Since this inconsistency affects the computed energy of the remaining material (and thus the determination of how much envelope mass has been unbound), it is very important to make the box large enough to contain as much of the envelope as possible throughout the simulation.

A final consideration is the manner in which the gravitational acceleration is coupled to the hydrodynamical equations. Depending on how this coupling is structured, it can have implications for the conservation of energy and momentum (Section 4.3.3). Additionally, if Equation (4.9) is represented using a second-order finite difference, as in most codes, the same equation can be used to solve for the value of ϕ at zone centers (in which case the source density ρ is also defined at zone centers) or the zone-averaged potential $\bar{\phi}$ (in which case the zone-averaged density $\bar{\rho}$ is the source). The latter interpretation is used by finite-volume codes. In codes based on Godunov-type methods, the zone-averaged acceleration $\bar{\mathbf{g}}$ is interpolated to zone edges and used to correct the input states for the Riemann problems from which fluxes are derived. Thus care must be taken when adding gravity to ensure that second-order accuracy is preserved.

In contrast to AMR, since moving-mesh methods inherit from smoothed particle hydrodynamics the notion of a scattered collection of mass points that represent the gas, they use algorithms based on the integral form of Poisson's equation to determine the gravitational acceleration. Thus AREPO and MANGA, for example, use a tree algorithm (described in more detail in Section 4.2.3) to compute $\mathbf{g}$. For the purpose of computing the gravitational acceleration and potential, the mesh cells are treated as uniform-density spheres with volumes slightly larger than the volumes of the cells. The gravitational terms in the Euler equations are then applied to the cells in a fashion that splits the effect of the motion of the cells from the effect due to the advection of material between cells (Springel 2010). This choice improves the conservation of total energy in self-gravitating flows.

4.2.3 Smoothed Particle Hydrodynamics

Smoothed particle hydrodynamics (SPH) adopts a Lagrangian particle-based approach to solving the hydrodynamical equations (Rosswog 2009; Price 2012). The method dates to the 1970s (Lucy 1977; Gingold & Monaghan 1977). SPH was among the first numerical techniques applied to 3D common envelope simulations (de Kool 1987; Terman et al. 1994; Rasio & Livio 1996), though the codes in use at that time did not consistently account for spatially varying resolution and the number of particles used was low. Modern SPH codes in current use for common envelope simulations include GADGET (Springel 2010; Pakmor et al. 2012), originally developed for cosmological simulations; StarSmasher (Gaburov et al. 2010), a descendant of Rasio's original SPH code StarCrash; and Phantom (Price et al. 2018).

In SPH, a collection of N particles with masses m_i and positions $\mathbf{x}_i$ is used to represent the gas density field via a kernel density sum. The density associated with the ith particle is given, for example, by

$$\rho_i = \sum_{j=1}^{N} m_j W(|\mathbf{x}_i - \mathbf{x}_j|, h_j). \tag{4.36}$$

The kernel function $W(r, h)$ depends on the interparticle separation r and a smoothing length h. A popular choice is the cubic spline kernel

$$W(r, h) = \frac{1}{\pi h^3} \begin{cases} \frac{1}{4}(2 - r/h)^3 - (1 - r/h)^3 & r \leqslant h \\ \frac{1}{4}(2 - r/h)^3 & h < r \leqslant 2h \\ 0 & r > 2h \end{cases}. \tag{4.37}$$

In modern implementations the smoothing length is allowed to vary inversely with the density, allowing higher resolution in higher-density regions. To accomplish this, at each time step Equation (4.36) is iteratively solved together with

$$h_i = \eta \left(\frac{m_i}{\rho_i} \right)^{1/3}, \tag{4.38}$$

where η is a parameter related to the mean number of neighbors each particle has within radius $2h_i$. For the cubic spline kernel, the mean number of neighbors in regions of constant density is $N_{\rm neigh} = \frac{32}{3}\pi\eta^3$. With too many close neighbors, SPH with a cubic spline kernel is subject to a weak pairing instability that can be avoided by using $\eta = 1.2$, which corresponds to $N_{\rm neigh} \approx 58$. In particular, with this kernel it is important not to increase the number of neighbors with increasing number of particles. However, keeping the number of neighbors fixed while increasing N may impede convergence, particularly in regions with large density gradients (Zhu et al. 2015). One way to simultaneously explore the limits $N \to \infty$, $N_{\text{neigh}} \to \infty$, $h \to 0$, while avoiding pairing instabilities, is to use the Wendland C^4 kernel (Wendland 1995), which is not subject to those instabilities (Dehnen & Aly 2012).

When h is allowed to vary, for consistency it is important to derive the equations of motion from a Lagrangian function (Nelson & Papaloizou 1994; Springel & Hernquist 2002). For the above definitions the appropriate Lagrangian is

$$L = \sum_{i=1}^{N} m_i \left[\frac{1}{2} v_i^2 - u_i(\rho_i, P_i) \right], \tag{4.39}$$

where $\mathbf{v}_i = d\mathbf{x}_i/dt$ is the ith particle's velocity, u_i is its specific internal energy, and P_i is its pressure. The corresponding equations of motion are

$$\frac{d\mathbf{v}_i}{dt} = -\sum_{j=1}^{N} m_j \left[\frac{P_i}{\Omega_i \rho_i^2} \frac{\partial W(r_{ij}, h_i)}{\partial \mathbf{x}_i} + \frac{P_j}{\Omega_j \rho_j^2} \frac{\partial W(r_{ij}, h_j)}{\partial \mathbf{x}_i} \right] \tag{4.40}$$

$$\frac{du_i}{dt} = \frac{P_i}{\Omega_i \rho_i^2} \sum_{j=1}^{N} m_j (\mathbf{v}_i - \mathbf{v}_j) \cdot \frac{\partial W(r_{ij}, h_i)}{\partial \mathbf{x}_i}, \tag{4.41}$$

where $r_{ij} = |\mathbf{x}_i - \mathbf{x}_j|$ and the quantity

$$\Omega_i \equiv 1 - \frac{\partial h_i}{\partial \rho_i} \sum_{j=1}^{N} m_j \frac{\partial W(r_{ij}, h_i)}{\partial h_i} \tag{4.42}$$

is a correction factor that accounts for the effect of varying smoothing lengths. With these equations of motion the total energy and momentum are conserved to the extent that the time integration scheme used is conservative. A second-order "kick–drift–kick" leapfrog method is often used to integrate the equations for $\mathbf{x}_i$ and $\mathbf{v}_i$. Since this method is symplectic, the total energy is conserved to machine precision. The internal energy Equation (4.41), when it is used, is integrated similarly to the velocity equation (e.g., in Phantom; Price et al. 2018).

The entropic variable $A_i = P_i \rho_i^{-\Gamma_{1,i}}$, is conserved in inviscid flow ($dA_i/dt = 0$), so some codes (e.g., later versions of StarSmasher; Lombardi et al. 2011) rely on it instead. For polytropic gases, A_i corresponds to the specific entropy. In codes that track this entropic variable, the specific internal energy can be computed via

$$u_i = \frac{A_i}{\Gamma_{3,i} - 1} \rho_i^{\Gamma_{1,i} - 1}. \tag{4.43}$$

Note that using this entropy formulation with a realistic equation of state requires that the adiabatic exponents be available as tabulated functions of density and pressure.

Dissipation and Mixing

The conservation of entropy is a problem for compressible flows in inviscid SPH since shock waves correspond to discontinuous jumps in the entropy. Recovering this behavior without affecting smooth flow requires introducing artificial viscosity with a switch that turns it on only in shocks. Artificial viscosity acts to smear out shocks so that they become resolvable, generating entropy in the process.

Modern SPH codes often use a form of artificial viscosity similar to that introduced by Monaghan (1997) using an analogy with the Riemann solvers used in grid-based codes. This approach requires dissipation terms to be added to both the velocity and internal energy (or entropy) equations. A dimensionless viscosity coefficient that multiplies these terms is stored for each particle. For example, in the formulation of Morris & Monaghan (1997) the viscosity coefficients obey a time evolution equation with a source term that is nonzero at compressive velocity jumps and an exponential sink term to produce damping away from shocks. Alternatively, the method of Cullen & Dehnen (2010) uses a functional expression to set the viscosity coefficients and applies it using a shock-capturing criterion based on the time derivative of the velocity divergence, which allows it to be applied just before a particle passes through a shock and nowhere else.

Because of the need for artificial viscosity and the lack of resolution in low-density regions, the treatment of shocks and flow discontinuities is much worse in SPH than in Godunov-based AMR methods. Since the latter use Riemann solvers to capture shocks, their intrinsic numerical dissipation is close to the minimum possible (corresponding to resolution of a discontinuity with 1–2 zones). PPM does introduce a small amount of artificial viscosity to damp oscillations behind very strong shocks, but in practice this should not even be triggered in common-envelope simulations.

SPH has particular difficulty with shear flows and contact discontinuities. This was vividly demonstrated by the "Wengen blob" code comparison (Agertz et al. 2007). In this test, a uniform, spherical gas blob moves supersonically with respect to a homogeneous medium initially in pressure equilibrium with the blob. Shear instabilities should rapidly destroy the blob, and indeed this was seen in the three AMR codes used for the test, but the two SPH codes strongly suppressed the growth of these instabilities, allowing the blob to persist (though strongly flattened by ram pressure).

Though it was anticipated by previous work (Morris 1996; Dilts 1999; Ritchie & Thomas 2001; Marri & White 2003), this test result led to a major re-evaluation of SPH in the community (and contributed to the motivation for the development of moving-mesh codes). For example, Price (2008) showed that Lagrangian-based SPH methods implicitly assume the differentiability of fluid quantities and thus neglect surface integral terms in the momentum and energy equations that appear at flow discontinuities. The neglect of these terms produces pressure errors at contact discontinuities that tend to drive apart particles on either side of the interface. As we have seen, SPH traditionally uses artificial viscosity to resolve shocks, which involve jumps in the pressure and velocity. However, Price showed that contact discontinuities, across which the pressure and velocity are constant, also must be regularized. Because it relies on the pressure and velocity jumps, artificial viscosity would not be triggered at such locations. An artificial *conductivity* must also be added; this acts on the jump in internal energy at contact discontinuities. With this change Price was able to reproduce the expected growth of Kelvin–Helmholtz instability at shear interfaces. Wadsley et al. (2008) made similar arguments but resolved the issue by adding heat diffusion based on a subresolution turbulence model to SPH.

Read et al. (2010) unified the different explanations for the problem of shear flows by performing a stability analysis of a generalized set of SPH equations. They showed that when particles' kernel functions are sampled by a finite number of neighbors, SPH is subject to instabilities that exacerbate errors in density gradients at discontinuities and prevent particles from mixing on scales smaller than the smoothing length. Part of the solution pursued by Read et al. (and similar to one adopted earlier by Ritchie & Thomas 2001) was to modify the kernel density estimator to rely on the entropic variable A_i and thereby produce smaller smoothing lengths at discontinuities. Interestingly, whereas the solution of Price (2008) and Wadsley et al. (2008) was to make SPH more dissipative at contact discontinuities, the solution of Ritchie & Thomas (2001) and Read et al. (2010) was effectively to represent entropy-generating mixing through a more stochastic particle distribution. The former approach depends more on the assumption of a subresolution model for mixing and entropy generation, while the latter produces a noisier particle distribution.

When comparing SPH simulation results to the output of grid-based codes, it is important to know which SPH approach was adopted and use the appropriate kernel sum definitions to compute fluid quantities. The different approaches may also require very different particle counts to reach a given level of convergence, particularly if flow discontinuities in low-density material are important.

Including Self-gravity in SPH

Since SPH employs stochastically distributed mass points to represent the gas, it is natural to solve the Poisson equation for the gravitational field using an integral rather than a differential representation. In other words, rather than solving the Poisson partial differential equation, to find the force on an SPH particle we sum the accelerations due to the rest of the SPH particles:

$$\mathbf{g}_{\mathrm{gas},i} = \sum_{j=1}^{N} \mathbf{g}_{ij} = -G\sum_{j=1}^{N} \frac{m_j}{|\mathbf{x}_i - \mathbf{x}_j|^3}(\mathbf{x}_i - \mathbf{x}_j). \tag{4.44}$$

When "core particles" are included for the core of the donor and the companion, the result of this sum can simply be added to the core-sourced accelerations as described in Section 4.2.2.

However, Equation (4.44) has two serious flaws. First, it is a very hard force law, so particles that approach each other will scatter in unphysically vigorous ways that efficiently damp bulk kinetic energy (a phenomenon known as two-body relaxation). Since in SPH the particles are meant to more-or-less smoothly represent the gas density field rather than track the motion of real particles, extreme collisions between these resolution elements would seem to be a bad thing. Second, since computing the acceleration of each particle requires a sum over all of the other particles, the cost of the calculation scales as the square of the number of particles. This rapidly becomes very expensive.

To solve the first problem, the interparticle force is *softened* to prevent pairs of particles from scattering through large angles if they pass close to each other. A popular choice is Plummer softening, for which

$$\mathbf{g}_{ij} = -\frac{Gm_j}{(|\mathbf{x}_i - \mathbf{x}_j|^2 + \varepsilon^2)^{3/2}}(\mathbf{x}_i - \mathbf{x}_j). \tag{4.45}$$

Here ε is referred to as the softening length, not to be confused—though understandably it often is—with the SPH smoothing length h. Further confusing matters, although ε and h are conceptually separate, in self-gravitating flows they should be numerically the same to avoid unphysical results (Bate & Burkert 1997). Additionally, using a symmetrized, softened force law based on the SPH kernel produces forces that are more consistent with the Poisson equation given the SPH densities as a source term (Dehnen 2001). Price & Monaghan (2007) show that when variable smoothing lengths are used with this approach, it is necessary to add an acceleration term involving factors analogous to Equation (4.42) to ensure energy conservation.

The second problem is resolved in SPH by using a hierarchical approximation method to limit the number of terms computed in Equation (4.44).[2] The most common approach is known as the hierarchical tree algorithm (Barnes & Hut 1986; Hernquist & Katz 1989). Simply put, the direct summation in Equation (4.44) is carried out only for particle pairs that are sufficiently close; the influence of more distant particles is included by aggregating them and using the monopole (or in some implementations, the monopole and quadrupole) force due to the aggregates. Since SPH requires a nearest-neighbor search anyway to compute kernel sums, and since such searches are easily accelerated using hierarchical trees, the tree algorithm conveniently exploits software infrastructure that already had to be built.

The tree algorithm is most efficient for highly clustered distributions of mass points, so it is most appropriate for the early stages of common-envelope evolution. Later, when the mass is more spread out, it should become less efficient, though still much better than direct summation. In cosmological simulations the hybrid "tree–particle-mesh" algorithm is used to smoothly blend a grid-based potential solver

[2] Some codes, such as StarSmasher (Gaburov et al. 2018), do employ direct summation, but carry it out on graphics processing units (GPUs) to make it computationally feasible.

where the matter distribution is more uniform with a tree-based solver where it becomes more clumpy (Xu 1995; Bode et al. 2000; Bagla 2002). This algorithm is available in some of the SPH codes used for common-envelope simulations (notably GADGET).

4.3 What Can We Trust?

Regardless of the numerical method or code used, or whether the initial conditions are correct, any hydrodynamical simulation code will be subject to different types of error. Moreover, no generally applicable error bounds are known to exist for numerical solutions of the Euler or Navier–Stokes equations. We must therefore live with rough and ready estimates of error constructed by solving problems with known solutions (*verification*), comparing simulations with experiments (*validation*) or other codes (*code comparison*), and studying solution behavior as spatial resolution or regularizing parameters are reduced to zero (*self-convergence*).

4.3.1 Truncation and Roundoff Error

First, an explanation of the main types of error is in order. We will only give a brief summary here; more detailed treatments are available in any competent text on numerical methods (e.g., LeVeque 1992).

Any numerical method for solving the hydrodynamical equations is subject to *truncation error* because it replaces a continuum partial differential equation (e.g., the continuity equation) with a system of algebraic difference equations. These may be obtained by replacing the derivatives with suitable difference approximations generated using Taylor expansions, as in the case of finite-difference methods. For example, if we denote by ρ_i the density at a position $x_i = i\Delta x$, where Δx is a uniform mesh spacing, then approximations to $\frac{\partial \rho}{\partial x}$ can be constructed using mesh point i and its neighbors by writing Taylor expansions for $\rho(x)$ about x_i at each point, truncating the expansions after a finite number of terms, and combining the expansions at several points to solve for the first derivative at x_i:

$$\left(\frac{\partial \rho}{\partial x}\right)_i \approx \frac{\rho_{i+1} - \rho_i}{\Delta x} + \mathcal{O}(\Delta x) \tag{4.46}$$

$$\approx \frac{\rho_{i+1} - \rho_{i-1}}{\Delta x} + \mathcal{O}(\Delta x^2) \tag{4.47}$$

$$\approx \frac{-\rho_{i+2} + 8\rho_{i+1} - 8\rho_{i-1} + \rho_{i-2}}{12\Delta x} + \mathcal{O}(\Delta x^4). \tag{4.48}$$

In each case, the truncation of the Taylor expansions produces terms proportional to Δx^p and higher powers of Δx for some integer p. Note that the "big-O" notation here does not have the same meaning as astrophysicists are accustomed to, where any multiplying coefficient is considered to be comparable to unity; instead $\mathcal{O}(\Delta x^p)$ should be read as "a polynomial whose leading term is proportional to Δx^p."

In particular, nothing is implied about the magnitude of the coefficient multiplying the leading term (we will return to this point later).

Truncation error also emerges in finite-volume, moving-mesh, and SPH methods; here what is evolved is not a set of density values, but rather integrals of the density multiplied by some weighting kernel over a resolution element which might be rectangular, tetrahedral, or constrained only by the kernel's support (as for SPH).[3] For example, consider finite-volume methods in which we average Equation (4.34) over a time interval $[t_n, t_n + \Delta t)$, producing the equation

$$\frac{\bar{\rho}_i^{n+1} - \bar{\rho}_i^n}{\Delta t} = -\frac{\overline{\rho v}_{i+1/2}^{n+1/2} - \overline{\rho v}_{i-1/2}^{n+1/2}}{\Delta x}. \tag{4.49}$$

This is an *exact* equation for the updated zone averages $\bar{\rho}_i^{n+1}$ given the averages at the previous time step and the fluxes on the right-hand side, which are *time-averaged* fluxes that are spatially averaged over the zone interfaces. Truncation error enters when these fluxes are replaced by an approximate expression, either through Taylor expansion, Riemann problem solution, or some other technique.

Generally speaking, the time step Δt and spatial resolution Δx cannot be chosen independently. Numerical methods for Equation (4.49) in which the approximate fluxes are constructed entirely from information at time t_n and earlier are referred to as *explicit*; *implicit* methods include information from time t_{n+1} and must be solved using linear algebra techniques. To avoid numerical instability, in which the solution becomes dominated by exponentially growing errors, explicit methods require some form of the Courant–Friedrichs–Lewy (CFL) time step constraint,

$$\Delta t \leqslant \sigma \min_i \left(\frac{\Delta x}{c_{s,i} + |v_i|} \right), \tag{4.50}$$

where σ is a coefficient smaller than one called the CFL parameter and the minimum is taken over all zones. Thus smaller resolution elements require smaller time steps. Implicit methods usually sidestep the CFL constraint, allowing larger time steps, but accuracy diminishes as the time step is increased. They are not often used for highly compressible flows.

A second major type of error, *roundoff error*, arises because of the finite representation of numbers by a computer. Even numbers like 0.1 that can be represented by a finite sequence of digits in decimal notation must be represented by truncated sequences of digits in binary. Thus most arithmetic operations on floating-point numbers will produce a result that is incorrect in the final digit. Sometimes these errors can be large, as when two very similar numbers are differenced; for highly supersonic flows this can lead to problems with the updated internal energy if it is determined by differencing the updated total and kinetic energies. Even when

[3] Strictly speaking, in most implementations of SPH the continuity equation is not directly solved. However, all quantities attached to SPH particles are regarded as kernel sums. SPH additionally differs from finite-volume and moving-mesh techniques in that its "resolution elements" are not disjoint spatial volumes; the kernels associated with particles overlap.

the roundoff error at each time step is small, it grows with each step, and evolving through a very large number of time steps (e.g., $>10^6$) can cause it to be significant. Thus truncation error and roundoff error exist in tension: increasing the resolution to reduce truncation error causes more steps to be taken (for stability) to reach a given simulation time, increasing the roundoff error.

To combat this tension, we can use double-precision arithmetic and higher-order numerical methods to reduce the truncation error for a given resolution. For many problems, second-order spatial truncation error is satisfactory, but for turbulent flows higher order is often desirable. If the order of the truncation error is even, it may resemble a numerical approximation to a diffusion operator, and we speak of *numerical diffusion* or *numerical viscosity*. If it is odd, the leading error terms may behave like a wave equation, and discontinuous flow will produce oscillatory errors. Generally the truncation error is some combination of these features, and high-order methods are constructed explicitly to limit the amount of oscillation without introducing too much diffusion.

Mixing is one important concern related to truncation error and numerical diffusion. Eulerian codes inherently mix adjacent regions of fluid within zones, because a sharp interface between two fluids that moves less than a zone width during a time step must be smeared to be at least one zone wide in order to be represented on the mesh. Along with this mixing of mass and composition comes mixing of energy and angular momentum. SPH and moving-mesh codes (the latter to the extent that the mesh moves with the gas) are Lagrangian and so no subresolution mixing takes place. This difference is not the same as that seen in the Wengen blob test discussed in Section 4.2.3; but it is connected. Ironically, reducing the numerical dissipation of Eulerian codes to better-represent inviscid gases causes them to very readily amplify instabilities seeded by numerical errors. This increases the surface area of contact between flow regions and allows them to (numerically) mix more efficiently at the zone level than Navier–Stokes solvers with explicit diffusion (Lecoanet et al. 2016). In this respect, Eulerian and Lagrangian codes bracket the likely true situation, with Eulerian codes overestimating the amount of mixing and Lagrangian ones suppressing it entirely. Since AMR and SPH codes seem to broadly agree when run on the same systems (Section 4.3.4), it is not clear what the implications are for common-envelope simulations. However, we might expect that 3D runs carried far into the slow spiral-in phase would show differing amounts of envelope ejection after a given time with Lagrangian codes than with Eulerian ones.

4.3.2 Convergence

For modern numerical methods the truncation error depends on the solution, so formal convergence (in the sense of a bound on the coefficients of the truncation error terms) is difficult to prove. For example, as the resolution is increased the effective numerical viscosity decreases, and beyond some limit the flow may become turbulent, producing a discontinuous change in the truncation error.

Simulations that include other types of physics such as radiation, magnetic fields, or reactions can experience other types of change in their convergence behavior as resolution is increased. Therefore it is essential for simulation practitioners to demonstrate at least a hacker's form of convergence by varying the resolution of their calculations and showing that some measure of the error (e.g., the L2 norm of some quantity,[4] the total unbound mass, etc.) indeed approaches a finite limit as resolution increases. Since the problem being solved does not have a known analytical solution, what we really attempt to do is to show "self-convergence" to the result of a simulation performed at high resolution.

Note that the gas resolution element size is not the only relevant numerical parameter for which convergence must be demonstrated. In SPH with varying kernel smoothing lengths, the number of particle neighbors or the total mass within a smoothing length is kept roughly fixed in regions with different densities. For a given total particle count, therefore, decreasing the smoothing length decreases the number of neighbors, which increases shot noise in the gas state quantities. Thus reducing the smoothing lengths requires increasing the particle count. But should the average number of neighbors per particle also be increased? As discussed in Section 4.2.3, formal convergence of SPH requires simultaneously taking the limits $N \to \infty$, $N_{\rm neigh} \to \infty$, $h \to 0$, but if the number of neighbors is increased with the particle count for traditional SPH kernel functions, the particles will tend to clump. The Wendland C^4 and similar kernels are not subject to this pairing instability, so the number of neighbors can be increased with the particle count, producing better convergence. However, increasing the number of neighbors increases the cost of computing kernel sums.

SPH particles additionally have a separate gravitational softening length that is used to reduce the effects of two-body relaxation. To avoid unphysical (de) stabilization of self-gravitating flows, this softening length should be taken to be equal to the SPH smoothing length (Bate & Burkert 1997). For mesh-based codes, the differencing of the potential on the mesh to determine the gravitational acceleration effectively ties the gravitational softening to the hydrodynamical resolution. If the gravitational acceleration is obtained by differencing the potential on a grid, even a point mass field will effectively be softened over about two resolution elements. Thus in both cases the effective gravitational softening length is 1–2× the hydrodynamical resolution element size. One should not expect self-gravitating flow features to be recovered properly below this scale (though this problem is more important in simulations of the interstellar medium than for common-envelope simulations, where the minimum size of gravitationally unstable perturbations is much bigger).

For both SPH- and mesh-based approaches the use of a numerical stellar core introduces an additional length scale into the problem. In most simulations these cores are treated as particles and not a special kind of internal boundary, so some gas remains within the core radius. Whether or not this gas can be maintained in

[4] The Ln error norm for a quantity Q_i defined on resolution elements i at positions x_i, in comparison with the correct solution $Q(x_i)$, is defined as $\left[\sum_i |Q_i - Q(x_i)|^n\right]^{1/n}$.

hydrostatic equilibrium by the code depends on the ratio of core radius to gas resolution element length. Ohlmann et al. (2016) suggest this ratio should be at least 20, while Sand et al. (2020) require a factor of 40 for an AGB donor (both with the moving-mesh code AREPO). Chamandy et al. (2019), using the AMR code AstroBEAR, study the effect of varying this ratio in the range 9–17 and find that a smaller value of the ratio produces a larger final separation and smaller eccentricity. Since gravity treatments vary significantly among different types of code, a careful inter-code comparison of these resolution criteria is still needed.

Key measures of convergence for the common-envelope problem include the final orbital separation and the total unbound mass. Convergence studies by Iaconi et al. (2017) for SPH and Iaconi et al. (2018) for AMR suggest that with increasing resolution the orbital separation decreases and the total unbound mass increases. Although these works simply report on results and do not convincingly explain this trend, one possibility discussed by them is that increasing the resolution in the vicinity of the companion increases the effective drag. We may speculate that if envelope material is able to swing closer to the companion with higher resolution,[5] it will apply a stronger local tidal drag on the companion. If this is so, adequately resolving the accretion flow around black hole or neutron star companions in particular may be very difficult. In contrast, the global tidal drag forces that are important at early times should be easier to compute accurately since they result from relatively low-order perturbations of the donor (e.g., MacLeod et al. 2019).

Another consideration is that the Iaconi et al. simulations were still a factor of ten or more in length scale from resolving the physical stellar core. In this regime, with increasing resolution the total resolved envelope binding energy steadily increases, making envelope removal more difficult without reaching a clear point at which the inspiral should naturally terminate. Presumably we should not expect to see convergence until the separation at which whatever physical criterion halts the inspiral is refined.

4.3.3 Conservation

In the absence of radiative losses, the equations of hydrodynamics are local conservation laws which, when integrated over all space, correspond to statements of the global conservation of mass, linear momentum, and energy (kinetic plus internal plus potential). Correct solutions to the continuum equations obey these global conservation laws. Is this property preserved by numerical solutions containing various forms of truncation error? Yes and no.

As discussed in Section 4.2.2, finite-volume AMR and moving-mesh techniques discretize the hydrodynamical equations in such a way that total mass, linear momentum, and kinetic plus internal energy are explicitly conserved.[6] Regardless of the accuracy of these fluxes—indeed, even if the fluxes are replaced by random

[5] Here we assume the ratio of the effective numerical core radius to the gas resolution element length is fixed.
[6] AMR methods require an additional "flux conservation" step at refinement boundaries to ensure that, e.g., the total mass flux exiting a coarse zone equals the sum of the mass fluxes entering its refined neighbors.

numbers—a properly written code with periodic or reflecting boundaries should conserve these three quantities to within roundoff error. With outflow or inflow boundaries the total change in each quantity should similarly match that expected from the total fluxes over the exterior boundary of the computational volume. Note that conserving the totals of these quantities does not guarantee the accuracy of the solution; that depends on the truncation error in the fluxes.

However, once gravity is introduced, this desirable state of affairs sadly no longer holds. The gravitational source terms ($-\rho\nabla\phi = \rho\mathbf{g}$ in the momentum equation and $-\rho\mathbf{v}\cdot\nabla\phi = \rho\mathbf{v}\cdot\mathbf{g}$ in the energy equation) are not divergences of fluxes, and for a general externally-imposed potential a mesh code cannot exactly conserve the total energy including the potential energy. However, for a self-gravitating system, in which the potential is obtained from the Poisson equation, it is possible to write

$$\rho\mathbf{g} = \nabla\cdot\mathbf{G}, \tag{4.51}$$

where

$$G_{ij} = -\frac{1}{4\pi G}\left(g_i g_j - \frac{1}{2}|\mathbf{g}|^2\delta_{ij}\right) \tag{4.52}$$

is the gravitational analog of the Maxwell stress tensor in electromagnetism. Thus even with self-gravity the momentum equation can be differenced in a conservative fashion. Not all codes implement this feature, however, and for these codes it is especially important to perform single-star or wide-binary advection tests (discussed later) to establish the level of linear momentum conservation. Since common-envelope simulations are performed with open boundaries through which material can be lost, it can be difficult to distinguish genuine nonconservation from the loss of momentum carried by outflowing material.

A similar approach can be used to conservatively difference the energy equation (Jiang et al. 2013). If the inviscid energy equation is written in the form

$$\frac{\partial}{\partial t}\left(\rho E + \frac{1}{2}\rho\phi\right) + \nabla\cdot\left[(\rho E + P)\mathbf{v} + \mathbf{F}_{\mathrm{g}}\right] = 0, \tag{4.53}$$

then using the Poisson equation it is possible to show that a (non-unique) choice for $\mathbf{F}_{\mathrm{g}}$ is

$$\mathbf{F}_{\mathrm{g}} = \frac{1}{8\pi G}(\phi\nabla\dot{\phi} - \dot{\phi}\nabla\phi) + \rho\mathbf{v}\phi. \tag{4.54}$$

Computing the time derivative of the potential to second-order accuracy requires an additional Poisson solver call (which can be a significant expense), so this approach to energy conservation has a much bigger code performance impact than differencing the stress tensor for the momentum equation. It is most important for marginally bound systems (and thus might be worth considering, e.g., for supergiant envelopes). However, most codes do not implement it.

Typical levels of total energy nonconservation for AMR/MM common-envelope simulations in the literature are at the percent level. For example, Ohlmann et al.

(2016) report 3% energy nonconservation over 80 orbits of a low-mass giant inspiral calculation using AREPO. Iaconi et al. (2018) report a decrease of less than 1% in the total energy in their Enzo AMR simulations over a similar number of orbits, but this change was a larger fraction (~10%) of the initial envelope binding energy. They additionally observed spikes in the potential energy of some zones early in the inspiral, which they attributed to AMR regridding artifacts but argued based on a slightly different calculation did not affect the final result. Chamandy et al. (2019) mention a 5% change in total energy over about ten orbits simulated with AstroBEAR. Energy nonconservation at these levels is not likely to be a concern unless these codes are run for many hundreds of orbits. However, it is not clear whether errors in the kinetic or internal energy are being partially masked by errors in the potential energy.

Perhaps the most significant conservation-related concern for grid codes is angular momentum conservation. Because they do not explicitly solve an equation for angular momentum (which would overconstrain the velocity field), the level of angular momentum nonconservation is unconstrained and must simply be monitored. In nested-grid and AMR codes, Sandquist et al. (1998) and Iaconi et al. (2018) both report angular momentum nonconservation at the 10% level, which is clearly inadequate for following deep inspirals after which the orbital angular momentum may be only a few percent of its initial value. This is one area where moving-mesh algorithms appear to have an advantage over AMR; Ohlmann et al. (2016) observe a change of less than 1% in angular momentum for their calculations with AREPO. It should be said, however, that broadly speaking both types of code give results similar to SPH (Iaconi et al. 2018).

In fact, where conservation is concerned, SPH appears to have a clear advantage over AMR codes (and to a lesser extent moving-mesh codes). When formulated using a Lagrangian, SPH without artificial viscosity in principle exactly conserves energy; errors creep in through the time integration method used. Most implementations use a symplectic, time-symmetrized integrator such as leapfrog, enabling total energy errors to be kept below 1%, as is typically observed (e.g., Reichardt et al. 2019). Additionally, because close particle interaction forces are computed in a pairwise fashion and are parallel to the lines separating particles, total angular momentum is also conserved to better than 1% in SPH codes.

4.3.4 Verification and Validation

Given that our problem has no analytic solution—and if it did, simulations would be unnecessary—how do we know our simulations are converging to the physically correct solution as we increase resolution? Respecting global conservation laws reduces the dimensionality of the space the simulation explores, but it does not constrain the numerical solution to be a good approximation of the true dynamics. In the absence of rigorous error bounds we must establish confidence in simulation results using verification and validation tests. *Verification* and *validation* are terms with precise definitions in the computational fluid dynamics community (American Institute of Aeronautics and Astronautics 1998); they are sometimes confused by astrophysicists, but they have very different purposes and implications. Verification involves establishing that a code solves the equations it is supposed to solve, whereas validation involves establishing that those solutions represent the real world.

More precisely, verification refers to the demonstration that a simulation code produces numerical solutions that are convergent and consistent with the underlying partial differential equations of interest. Typically, in order to verify a code, we must show that it correctly solves (in the limit of infinite resolution) test problems having known solutions. Of course, to have an analytical solution a test problem must generally involve simplified physics, e.g., an ideal-gas equation of state instead of a tabulated one. Each of the code papers referenced in Table 4.1 presents a number of verification test results that exercise the code on problems such as 1D shock tubes (e.g., Sod 1978), spherically symmetric blast waves and implosions (Sedov 1959; Noh 1987), and simple advection of hydrostatic objects. The last of these can be a surprisingly difficult test to pass. An advected hydrostatic object (e.g., a polytrope or a star) is simply translated across the simulation domain and should retain its shape and density profile. In the process of moving across a grid, for example, oscillations about hydrostatic equilibrium are established by truncation errors and may not damp efficiently. In the common-envelope context, evolving a wide (non-interacting) binary is a related test that every code should pass.

Since verification tests generally do not exercise the full complexity of the physics included in many codes, it is crucial that a variety of tests be performed to exercise the code on as many combinations of physics units (e.g., hydrodynamics plus self-gravity, hydrodynamics plus radiation diffusion) as possible. These must also explore a range of different conditions; for example, in addition to the Sod shock-tube test, strong shock and rarefaction tests should be done. For codes under active development, it is best to automate verification testing to ensure that tests continue to be passed (e.g., FlashTest; Calder et al. 2006). In particular, users of publicly available codes should keep in mind that the version of a code they are using for a simulation may not be the same as the version that passed these basic tests in the original code paper. Users should produce their own verification tests (e.g., of the wide binary problem) both to demonstrate that the code is working for problems similar to the target problem and to establish convergence criteria.

Validation is a much more difficult problem. Strictly speaking, it refers to the comparison of a simulation code with physical experiments that demonstrate that the verified solutions produced by the code actually represent the real world. In astrophysics it is very difficult to set up reasonable validation experiments, though basic fluid instabilities are a particular area in which validation experiments can be done (e.g., Calder et al. 2002). Ultimately, we can construct synthetic observations from simulation results and compare them with astronomical observations, but the resulting comparisons are often fairly coarse-grained and admit multiple interpretations regarding what physics may be missing from the simulations. In addition, as discussed in Section 4.1.6, generating accurate light curves and colors from common-envelope simulations is still a challenge, one that must be solved if we are to exploit transient observations to constrain common-envelope physics.

Table 4.2. Simplified Comparison of Performance of the Main Types of Simulation Code

Physics	AMR	SPH	MM
Energy conservation	Worst	Best	Worst
Angular momentum conservation	Worst	Best	Intermediate
Galilean invariance	Worst	Best	Better
Advection error	Most	Least	Intermediate
Accurate gravity coupling	Tricky	Easy	Tricky
Vacuum boundary	No, "fluff"	Yes	Yes
Problem geometry	Hardest	Easiest	Easiest
Shock capturing	Best	Worst	Best
Shear instability capturing	Better	Suppressed (old codes), good but slow (modern codes)	Best
Artificial viscosity	Minimal	Significant in shocks	Minimal
Subresolution mixing	Most	Absent by default, but can be added (Read et al. 2010)	Less
Resolution in low-density regions	Easy	No	Easy
Resolution in high-density regions	Harder	Easier	Harder
Radiation hydro	Easiest	Hardest	Intermediate
Magnetohydrodynamics	Easiest	Hardest, but realizations exist (e.g., Price 2012)	Intermediate

In the absence of true validation tests, many simulation subfields in astrophysics have turned to code comparison projects. A classic of the genre is the Santa Barbara Cluster Comparison Project (Frenk et al. 1999), used to compare cosmological simulation codes. These projects are most valuable when they cover a range of algorithmic approaches and use a common analysis of the codes' output rather than allowing each group to perform a separate analysis. It must always be borne in mind that code comparison projects are not validation tests, as the different codes may jointly converge to the same wrong solution. However, they are important for establishing confidence and interpreting differences in the results obtained with different approaches.[7] As yet no code comparison project has been performed in the common-envelope simulation community, though Iaconi et al. (2017) compared results obtained with Enzo (AMR), Phantom (SPH), and SNSPH, finding broadly similar agreement (at the 10% level) for the amount of unbound mass, the timing of unbinding, and the final separation.

When interpreting simulation results it is important to keep in mind the different advantages and disadvantages of the numerical approaches in use. We conclude this Chapter by summarizing some of these differences in Table 4.2.

[7] They also very effectively expose bugs.

References

Agertz, O., Moore, B., Stadel, J., et al. 2007, MNRAS, 380, 963
American Institute of Aeronautics and Astronautics 1998, AIAA Guide for the Verification and Validation of Computational Fluid Dynamics Simulations, AIAA G-077-1998(2002) (Reston, VA: American Institute of Aeronautics and Astronautics), doi:10.2514/4.472855.001
Aspden, A., Nikiforakis, N., Dalziel, S., & Bell, J. 2008, Commun. Appl. Math. Comput. Sci., 3, 103
Bagla, J. S. 2002, JApA, 23, 185
Barnes, J., & Hut, P. 1986, Natur, 324, 446
Bate, M. R., & Burkert, A. 1997, MNRAS, 288, 1060
Berger, M. J., & Colella, P. 1989, JCoPh, 82, 64
Berger, M. J., & Oliger, J. 1984, JCoPh, 53, 484
Black, D. C., & Bodenheimer, P. 1975, ApJ, 199, 619
Bode, P., Ostriker, J. P., & Xu, G. 2000, ApJS, 128, 561
Brandt, A. 1977, MaCom, 31, 333
Bryan, G. L. 1996, PhD thesis, Univ. of Illinois at Urbana-Champaign
Bryan, G. L., Norman, M. L., O'Shea, B. W., et al. 2014, ApJS, 211, 19
Burkert, A., & Bodenheimer, P. 1993, MNRAS, 264, 798
Calder, A. C., Fryxell, B., Plewa, T., et al. 2002, ApJS, 143, 201
Calder, A. C., Taylor, N. T., Antypas, K., Sheeler, D., & Dubey, A. 2006, in ASP Conf. Ser. 359, Numerical Modeling of Space Plasma Flows: Astronum-2006, ed. N. V. Pogorelov, & G. P. Zank (San Francisco, CA: ASP), 119
Chamandy, L., Frank, A., Blackman, E. G., et al. 2018, MNRAS, 480, 1898
Chamandy, L., Tu, Y., Blackman, E. G., et al. 2019, MNRAS, 486, 1070
Chang, P., Wadsley, J., & Quinn, T. R. 2017, MNRAS, 471, 3577
Chen, K.-J., Heger, A., & Almgren, A. S. 2011, CoPhC, 182, 254
Chen, Z., Coleman, M. S. B., Blackman, E. G., & Frank, A. 2019, JCoPh, 388, 490
Colella, P., & Glaz, H. M. 1985, JCoPh, 59, 264
Colella, P., & Woodward, P. R. 1984, JCoPh, 54, 174
Cullen, L., & Dehnen, W. 2010, MNRAS, 408, 669
Cunningham, A. J., Frank, A., Varnière, P., Mitran, S., & Jones, T. W. 2009, ApJS, 182, 519
de Kool, M. 1987, PhD thesis, Univ. Amsterdam
Dehnen, W. 2001, MNRAS, 324, 273
Dehnen, W., & Aly, H. 2012, MNRAS, 425, 1068
Diehl, S., Rockefeller, G., Fryer, C. L., Riethmiller, D., & Statler, T. S. 2015, PASA, 32, e048
Dilts, G. A. 1999, IJNME, 44, 1115
Eggleton, P. P. 1971, MNRAS, 151, 351
Frenk, C. S., White, S. D. M., Bode, P., et al. 1999, ApJ, 525, 554
Fryer, C. L., Rockefeller, G., & Warren, M. S. 2006, ApJ, 643, 292
Fryxell, B., Olson, K., Ricker, P., et al. 2000, ApJS, 131, 273
Gaburov, E., Lombardi, J., James, C., Portegies Zwart, S., & Rasio, F. A. 2018, StarSmasher: Smoothed Particle Hydrodynamics code for smashing stars and planets, ascl:1805.010
Gaburov, E., Lombardi, J. C. Jr, & Portegies Zwart, S. 2010, MNRAS, 402, 105
Galaviz, P., De Marco, O., Passy, J.-C., Staff, J. E., & Iaconi, R. 2017, ApJS, 229, 36
Gingold, R. A., & Monaghan, J. J. 1977, MNRAS, 181, 375
Godunov, S. K. 1959, Mat. Sb., 47, 357

Hernquist, L., & Katz, N. 1989, ApJS, 70, 419
Hou, T. Y., & Floch, P. G. L. 1994, MaCom, 62, 497
Iaconi, R., De Marco, O., Passy, J.-C., & Staff, J. 2018, MNRAS, 477, 2349
Iaconi, R., Reichardt, T., Staff, J., et al. 2017, MNRAS, 464, 4028
Iglesias, C. A., & Rogers, F. J. 1996, ApJ, 464, 943
James, R. A. 1977, JCoPh, 25, 71
Jiang, Y.-F. F., Belyaev, M., Goodman, J., & Stone, J. M. 2013, NewA, 19, 48
Joyce, M., Lairmore, L., Price, D. J., Mohamed, S., & Reichardt, T. 2019, ApJ, 882, 63
Krumholz, M. R., McKee, C. F., & Klein, R. I. 2004, ApJ, 611, 399
Lecoanet, D., McCourt, M., Quataert, E., et al. 2016, MNRAS, 455, 4274
LeVeque, R. J. 1992, Numerical Methods for Conservation Laws, Lectures in Mathematics (2nd ed.; Berlin: Birkhäuser)
Levermore, C. D., & Pomraning, G. C. 1981, ApJ, 248, 321
Livio, M., & Soker, N. 1988, ApJ, 329, 764
Lombardi, J. C. Jr, Holtzman, W., Dooley, K. L., et al. 2011, ApJ, 737, 49
López-Cámara, D., De Colle, F., & Moreno Méndez, E. 2019, MNRAS, 482, 3646
Lucy, L. B. 1977, AJ, 82, 1013
MacLeod, M., Ostriker, E. C., & Stone, J. M. 2018, ApJ, 863, 5
MacLeod, M., & Ramirez-Ruiz, E. 2015, ApJ, 803, 41
MacLeod, M., Vick, M., Lai, D., & Stone, J. M. 2019, ApJ, 877, 28
Maeder, A., & Meynet, G. 2000, ARA&A, 38, 143
Marri, S., & White, S. D. M. 2003, MNRAS, 345, 561
Meynet, G., & Maeder, A. 1997, A&A, 321, 465
Mihalas, D., & Weibel Mihalas, B. 1984, Foundations of Radiation Hydrodynamics (New York: Oxford Univ. Press)
Monaghan, J. J. 1997, JCoPh, 136, 298
Morris, J. P. 1996, PASA, 13, 97
Morris, J. P., & Monaghan, J. J. 1997, JCoPh, 136, 41
Nelson, R. P., & Papaloizou, J. C. B. 1994, MNRAS, 270, 1
Noh, W. F. 1987, JCoPh, 72, 78
Ohlmann, S. T., Röpke, F. K., Pakmor, R., & Springel, V. 2016, ApJL, 816, L9
Ohlmann, S. T., Röpke, F. K., Pakmor, R., & Springel, V. 2017, A&A, 599, A5
O'Shea, B. W., Bryan, G., Bordner, J., et al. 2004, in Springer Lecture Notes in Computational Science and Engineering, Adaptive Mesh Refinement—Theory and Applications, ed. T. Plewa, T. Linde, & V. G. Weirs (Berlin: Springer), 341
Pakmor, R., Edelmann, P., Röpke, F. K., & Hillebrandt, W. 2012, MNRAS, 424, 2222
Passy, J.-C., De Marco, O., Fryer, C. L., et al. 2012, ApJ, 744, 52
Paxton, B., Bildsten, L., Dotter, A., et al. 2011, ApJS, 192, 3
Paxton, B., Cantiello, M., Arras, P., et al. 2013, ApJS, 208, 4
Paxton, B., Marchant, P., Schwab, J., et al. 2015, ApJS, 220, 15
Paxton, B., Schwab, J., Bauer, E. B., et al. 2018, ApJS, 234, 34
Paxton, B., Smolec, R., Schwab, J., et al. 2019, ApJS, 243, 10
Price, D. J. 2008, JCoPh, 227, 10040
Price, D. J. 2012, JCoPh, 231, 759
Price, D. J., & Monaghan, J. J. 2007, MNRAS, 374, 1347
Price, D. J., Wurster, J., Tricco, T. S., et al. 2018, PASA, 35, e031

Prust, L. J., & Chang, P. 2019, MNRAS, 486, 5809
Quirk, J. J. 1991, PhD thesis, Cranfield Inst. of Technology
Rasio, F. A., & Livio, M. 1996, ApJ, 471, 366
Read, J. I., Hayfield, T., & Agertz, O. 2010, MNRAS, 405, 1513
Reichardt, T. A., De Marco, O., Iaconi, R., Tout, C. A., & Price, D. J. 2019, MNRAS, 484, 631
Ricker, P. M., & Taam, R. E. 2008, ApJ, 672, L41
Ricker, P. M., & Taam, R. E. 2012, ApJ, 746, 74
Ricker, P. M., Timmes, F. X., Taam, R. E., & Webbink, R. F. 2019, in Proc. IAU Symp. 346, High-mass X-ray Binaries: Illuminating the Passage from Massive Binaries to Merging Compact Objects, ed. L. M. Oskinova, et al. (Cambridge: Cambridge Univ. Press), 449
Ritchie, B. W., & Thomas, P. A. 2001, MNRAS, 323, 743
Robertson, B. E., Kravtsov, A. V., Gnedin, N. Y., Abel, T., & Rudd, D. H. 2010, MNRAS, 401, 2463
Rosswog, S. 2009, NewAR, 53, 78
Rozyczka, M. 1985, A&A, 143, 59
Sand, C., Ohlmann, S. T., Schneider, F. R. N., Pakmor, R., & Roepke, F. K. 2020, arXiv:2007.11000
Sandquist, E. L., Taam, R. E., Chen, X., Bodenheimer, P., & Burkert, A. 1998, ApJ, 500, 909
Sedov, L. I. 1959, Similarity and Dimensional Methods in Mechanics (New York: Academic)
Shiber, S. 2018, Galax, 6, 96
Sod, G. A. 1978, JCoPh, 27, 1
Springel, V. 2005, MNRAS, 364, 1105
Springel, V. 2010, MNRAS, 401, 791
Springel, V. 2010, ARA&A, 48, 391
Springel, V., & Hernquist, L. 2002, MNRAS, 664, 649
Stone, J. M., Tomida, K., White, C. J., & Felker, K. G. 2020, ApJS, 249, 4
Sytine, I. V., Porter, D. H., Woodward, P. R., Hodson, S. W., & Winkler, K.-H. 2000, JCoPh, 158, 225
Taam, R. E., & Bodenheimer, P. 1989, ApJ, 337, 849
Terman, J. L., Taam, R. E., & Hernquist, L. 1994, ApJ, 422, 729
Toro, E. F., Spruce, M., & Speares, W. 1994, ShWav, 4, 25
Trottenberg, U., Oosterlee, C. W., & Schüller, A. 2001, Multigrid, Texts in Applied Mathematics. Bd., Vol. 33 (San Diego, CA: Academic)
van Leer, B. 1979, JCoPh, 32, 101
Wadsley, J. W., Veeravalli, G., & Couchman, H. M. P. 2008, MNRAS, 387, 427
Wendland, H. 1995, Adv. Comput. Math., 4, 389
Xu, G. 1995, ApJS, 98, 355
Yorke, H. W., Bodenheimer, P., & Taam, R. E. 1995, ApJ, 451, 308
Zhu, Q., Hernquist, L., & Li, Y. 2015, ApJ, 800, 6

Natalia Ivanova, Stephen Justham and Paul Ricker

Chapter 5

The Onset of the Common Envelope

The onset of the dynamical plunge-in phase is not expected to be completely sudden in all cases. The onset includes both the pre-Roche-lobe overflow evolution of the donor, during which the donor becomes more and more perturbed and may experience enhanced mass loss, and the period during which the ongoing Roche-lobe mass transfer becomes unstable. The donor at the start of the dynamical plunge may differ strongly from an unperturbed star of the same initial mass and the same age; in particular, it may have a very different mass. The timescale of the pre-common-envelope evolution (CEE), and the ability to remove much of the donor's mass before the CEE, also depends on how well the binary system is synchronized prior to the CEE.

5.1 Tides and Pre-CEE

The rate of tidal synchronization in pre-CEE can be estimated using Zahn's theory of tidal spin interactions (Zahn 1975, 1977, 1989, 2008). In that theory, a binary system evolves to a state of minimum kinetic energy, in which the orbit is circular and the component spins are both aligned and synchronized with the orbital motion, while it conserves angular momentum. The timescale for this tidal synchronization depends on how the kinetic energy was dissipated. If our donor has an outer convection zone, the dominant mechanism is viscous dissipation acting on the equilibrium tide. If the donor has an outer radiative zone, the dominant mechanism is radiative damping acting on the dynamical tide. (For a review, see Zahn 2008.) In what follows we will mainly refer to the equilibrium tide, as most pre-CE donors would have an outer convection zone.

The tidal torque $\mathcal{T}_{\rm TT}$ is (Zahn 2008)

$$\mathcal{T}_{\rm TT} = -\frac{GM_{\rm comp}^2}{R_{\rm d}}\left(\frac{R_{\rm d}}{a}\right)^6 \sin\alpha, \tag{5.1}$$

doi:10.1088/2514-3433/abb6f0ch5

where α is the tidal lag angle, i.e., the angle between the tidal bulge and the line between the centers of mass of the stars. A weak friction approximation is usually used to determine $\sin\alpha$. In that approximation, if the rotation rate of the donor is $\Omega_{\rm rot}$, the orbital angular velocity is $\Omega_{\rm orb}a$, and $t_{\rm diss}$ is the timescale on which kinetic energy dissipates, then

$$\mathcal{T}_{\rm TT} = -\frac{\Omega_{\rm rot} - \Omega_{\rm orb}}{t_{\rm diss}} \frac{M_{\rm comp}R_{\rm d}^2}{q^2}\left(\frac{R_{\rm d}}{a}\right)^6. \tag{5.2}$$

Here $q = M_{\rm d}/M_{\rm comp}$ as was defined previously in Chapter 2 (note that in Zahn (2008) the mass ratio is defined as $M_{\rm comp}/M_{\rm d}$). Equation (5.2) can be used as a first-order approximation for convective donors if one uses $t_{\rm diss} = \tau_{\rm conv}/6\lambda_2$, where $\tau_{\rm conv}$ is the global convective turn-over time, $\lambda_2 = 0.019\alpha_{\rm ml}^{4/3}$ is a unitless constant, and $\alpha_{\rm ml}$ is the mixing-length parameter. For a numerically stable approximation in the case of radiative donors, one can use the expression for the torque derived from Zahn's theory by Kushnir et al. (2017). The characteristic dissipation timescale for radiative damping is much longer than in the case of turbulent convection.

If $k_{\rm gyr}$ is the donor's radius of gyration, its moment of inertia is

$$I_{\rm d} = k_{\rm gyr}^2 M_{\rm d} R_{\rm d}^2. \tag{5.3}$$

For a low-mass giant, the radius of gyration usually satisfies $k_{\rm gyr}^2 \approx 0.1$ and decreases with increasing mass of the star (see e.g., Claret 2004, for $k_{\rm gyr}$). Combining the above, the synchronization timescale can be estimated as

$$\tau_{\rm sync} \approx t_{\rm diss}\frac{\Omega_{\rm rot}}{|\Omega_{\rm rot} - \Omega_{\rm orb}|}k_{\rm gyr}^2 q^2\left(\frac{a}{R_{\rm d}}\right)^6. \tag{5.4}$$

It is clear that, due to the strong power dependence of the ratio of the donor's radius to the orbital separation, tidal interactions become significant only when the donor's radius is comparable to the orbital separation. Whether the donor is spun up tidally or not depends on the timescale, $\tau_{\rm ev}$, at which it expands on its evolutionary timescale while its radius is comparable to the orbital separation; a rapidly expanding donor may start a CEE asynchronized.

Since the angular momentum of the orbit is usually much larger than the angular momentum stored in the two companions, the circularization timescale is longer than the synchronization timescale; it scales as $e^{-2}(a/R_{\rm d})^8$, where e is the eccentricity. So high eccentricities are relatively rapidly damped compared to moderate and low eccentricities. Note that if the progenitor binary was initially eccentric and $\tau_{\rm sync} > \tau_{\rm ev}$, the mass transfer would start with an eccentric orbit. This may occur for "early giants" which, prior to approaching the Roche lobe, had a primarily radiative envelope. It may also be significant for CEEs in binary systems formed in globular clusters via dynamical interactions. To our knowledge, there are no published studies of mass-transfer stability for eccentric orbits; population-synthesis studies thus necessarily assume stability criteria calculated for circular orbits even when the orbits are eccentric.

5.2 Darwin Instability

A CEE can have a rapid onset if the system is prone to Darwin instability, also known as secular tidal instability (Darwin 1879). This instability may take place if the system evolves into, or maintains, a synchronized state. While the donor evolution expands and increases its moment of inertia, its $\Omega_{\rm rot}$ is decreasing. To spin-up again to $\Omega_{\rm orb}$, the giant withdraws angular momentum from the orbit. Decrease of the orbital angular momentum causes the orbit to shrink and the orbital period to decrease. Then the tidal locking forces would try to further extract even more angular momentum from the orbit in order to remain synchronized. If the orbital moment of inertia $I_{\rm orb}$ is less than three times $I_{\rm d}$, the donor's moment of inertia ($I_{\rm orb} < 3I_{\rm d} = 3k_{\rm gyr}^2 M_{\rm d} R_{\rm d}^2$), then the runaway occurs with the orbital separation shrinking until the two stars are no longer two isolated objects (Hut 1980). Using the typical value $k_{\rm gyr}^2 = 0.1$, it can be shown that the Darwin instability, in a synchronous binary, takes place for binaries in which the mass ratio between the donor mass and the companion mass is large, $q \lesssim 9$.

For the Darwin instability to operate, $\tau_{\rm sync} < \tau_{\rm ev}$ is required. If this condition is not met, then other processes, for example wind mass loss, may alter the tidal orbital shrinkage and prevent the onset of the instability (see, for example, the discussion in Bear & Soker 2010). The binaries in which the Darwin instability would most likely take place contain donors with convective envelopes. The orbital shrinkage due to Darwin instability takes place on an ever decreasing timescale, and just before contact this timescale approaches the donor's dynamical timescale. As an example, the Darwin instability has been suggested as a potential mechanism driving the V1309 Sco binary merger event (Tylenda et al. 2011).

5.3 Onset Induced by a Tertiary Companion

If a binary is the inner binary of a triple system, angular momentum can be exchanged between the outer and inner orbits. This can cause the angle between the inner and outer orbits, i.e., the mutual inclination angle, to oscillate along with the orbital eccentricities. When the timescale over which the orbits change is long compared to both orbital periods then this can be analyzed using an orbit-averaged secular approximation. This was analytically explored up to the quadrupole term of the approximation by Lidov (1962) and Kozai (1962) for solar system objects. Both Lidov and Kozai treated one component of the inner binary as a test particle, with negligible mass, in which case the Kozai–Lidov quadrupole oscillation cycles only occur when the mutual inclination angle between the inner and outer orbits is greater than approximately 40°.

Early work applying three-body dynamics to triple-star systems, also in the orbit-averaged secular approximation, included explicit mention of the octupole term (Harrington 1968, 1969), with the Kozai–Lidov quadrupole solution as a special case. Yet it later became fairly common to incorrectly refer to potential secular effects from a tertiary star on an inner binary as generically due to the Kozai–Lidov mechanism (often now Lidov–Kozai, since Lidov published first). However, there has been increasing recognition that the Kozai–Lidov approximations can lead to qualitatively misleading predictions for triple-star dynamics (see, e.g., Ford et al. 2000, or Naoz 2016 for an extensive modern review). For example, if we no longer restrict ourselves to the test-particle and quadrupole approximations, secular effects can become significant when the inner and outer orbits are much less misaligned than the 40° quoted for the Kozai–Lidov approximation.

Overall, the presence of a tertiary companion can cause the eccentricity of the inner orbit to increase, which may in turn lead to prompt mass loss during the pericenter passage and then the start of a CE phase. In such a situation any orbit-averaged approximation for the three-body dynamics would, of course, break down.

It is not guaranteed that a triple system stays bound through a common-envelope phase. When a significant fraction of the initial mass of the system is lost rapidly, the tertiary and the inner binary can unbind from each other (Michaely & Perets 2019). Nonetheless, some post-common-envelope binaries are observed to have wide tertiary companions, e.g., Wolf 1130 (Mace et al. 2018) and GD 319 (Farihi et al. 2005).

5.4 Orbital Evolution Due to Mass Loss

Orbital evolution due to mass loss or mass transfer cannot be treated using an energy formalism, as the mass ejection does not use only the orbital energy. Instead, the evolution is usually understood in terms of angular momentum conservation.

To understand the generic trend of the orbital behavior, let us consider a simple case where the donor can lose mass in the form of a wind or by transferring its mass to the companion, while the companion does not lose its own mass in a wind but can accrete a fraction β of the mass transferred by the donor. The relationship between the rates of change of the donor mass ($\dot{M}_\mathrm{d}$) and the companion mass ($\dot{M}_\mathrm{comp}$) is then

$$\dot{M}_\mathrm{comp} = -\beta \dot{M}_\mathrm{d}, \tag{5.5}$$

and the rate of mass loss from the binary system is

$$\dot{M}_\mathrm{d} + \dot{M}_\mathrm{comp} = (1 - \beta)\dot{M}_\mathrm{d}. \tag{5.6}$$

The total binary angular momentum can be written $J = J_\mathrm{orb} + J_\mathrm{d} + J_\mathrm{comp}$, where the three terms after the equals sign are, respectively, the angular momentum of the orbit about the system center of mass and the spins of the donor star and companion star about their own centers of mass. When the donor's radius approaches the size of the Roche lobe, its angular momentum J_d may be non-negligible when calculating the total system angular momentum. If the donor star rotates synchronously with the binary,

$$J_\mathrm{d} = \Omega_\mathrm{orb} I_\mathrm{d} = \Omega_\mathrm{orb} a^2 k_\mathrm{gyr}^2 r_\mathrm{d,l}^2 M_\mathrm{d}. \tag{5.7}$$

Here $r_\mathrm{d,l}$ is the effective Roche-lobe radius of the donor (see Equation (2.7)). For a 4:1 mass ratio between the donor and companion, at Roche-lobe overflow (RLOF) we can estimate that $J_\mathrm{d}/J_\mathrm{orb} \approx 0.12$. For simplicity in the following we consider an approximation in which the angular momenta of the donor and companion are neglected in comparison with the orbital angular momentum. For detailed calculations it is preferable to include the component stars' spin angular momenta, and in numerical simulations they are commonly taken into account.

Let the rate of angular-momentum loss from the binary due to mass loss be $\dot{J}_{\rm ML}$. The specific angular momentum $j_{\rm ML}$ carried away by the material lost from the binary can be parameterized as γ times the specific orbital angular momentum:

$$j_{\rm ML} \equiv \frac{\dot{J}_{\rm ML}}{\dot{M}_{\rm d} + \dot{M}_{\rm comp}} = \gamma \frac{J_{\rm orb}}{M_{\rm d} + M_{\rm comp}}. \tag{5.8}$$

$\dot{J}_{\rm ML}$ includes only the angular momentum carried away with the mass. In cases when other forms of angular-momentum loss need to be considered, for example due to magnetic braking or gravitational waves, one has to consider the total rate of change of the total orbital momentum $\dot{J} = \dot{J}_{\rm ML} + \dot{J}_{\rm MB} + \dot{J}_{\rm gw}$, and use on the LHS of Equation (5.8) $\dot{J}/(\dot{M}_{\rm d} + \dot{M}_{\rm comp})$.

The orbital angular momentum of an eccentric binary is

$$J_{\rm orb} = \sqrt{G} \frac{M_{\rm d} M_{\rm comp}}{\sqrt{M_{\rm d} + M_{\rm comp}}} \sqrt{a(1 - e^2)}. \tag{5.9}$$

We can differentiate this expression to obtain a general equation for the orbital evolution:

$$\frac{\dot{J}_{\rm orb}}{J_{\rm orb}} = \frac{1}{2}\frac{\dot{a}}{a} + \frac{\dot{M}_{\rm d}}{M_{\rm d}} + \frac{\dot{M}_{\rm comp}}{M_{\rm comp}} - \frac{1}{2}\frac{\dot{M}_{\rm d} + \dot{M}_{\rm comp}}{M_{\rm d} + M_{\rm comp}} - \frac{e\dot{e}}{1 - e^2}. \tag{5.10}$$

If the binary we consider has circularized (which does not have to be true, as this depends on the circularization timescale of the binary), or if $\dot{e}$ is very small (i.e., if the eccentricity evolves on a much longer timescale than mass transfer), the last term is zero. If we make the simplifying assumption that the last term can indeed be neglected, and assume $\dot{J}_{\rm orb} = \dot{J}_{\rm ML}$, then combine with the β and γ parameterizations above, we arrive at a parameterized expression for the orbital evolution due to the mass evolution of the binary:

$$\frac{\dot{a}}{a} = -2\frac{\dot{M}_{\rm d}}{M_{\rm d}}\left[1 - \beta\frac{M_{\rm d}}{M_{\rm comp}} - (1 - \beta)\left(\gamma + \frac{1}{2}\right)\frac{M_{\rm d}}{M_{\rm d} + M_{\rm comp}}\right]. \tag{5.11}$$

5.5 Increased Mass Loss Before the RLOF

A star in a binary that is close to RLOF may lose a significant amount of its mass well before the start of the mass transfer through the inner Lagrangian point. Below we list some of the proposed ways in which mass may be lost in close binaries just before the RLOF. All of the following suggested binary-enhanced mechanisms are speculative, even if inspired by observations, and some subsets of the list below may be attempts to describe the same phenomenology.

- **Tidally enhanced winds.** Tidal effects change the surface gravity of the donor and thus the rate of mass loss from its surface. A simple proposed model modifies an existing mass-loss prescription by multiplying by an ad hoc tidal-enhancement factor. For example, Tout & Eggleton (1988) proposed a multiplier for Reimers' mass loss rate $\dot{M}_R$ (Reimers 1975) that uses the scaling expected from Zahn's tidal theory, comparing the size of the donor to its effective Roche-lobe radius $R_{\rm d,L}$, along with a "boost factor" B:

$$\dot{M}_{\rm tid} = \dot{M}_R \times \left(1 + B \times \min\left[\left(\frac{R_{\rm d}}{R_{\rm d,L}}\right)^6, \frac{1}{2^6}\right]\right) \tag{5.12}$$

  The authors suggested a value for B of $\sim 10^4$. This value implies that at the moment of RLOF the wind mass loss rate is about 150 times more effective than the Reimers wind mass loss rate predicts. Initially the tidally-enhanced wind was proposed to explain the observed mass-ratio inversion in some RS CVn binaries. It has since been applied more broadly, e.g., for the symbiotic channel for Type Ia supernova, or to the horizontal branch morphology in globular clusters (Chen et al. 2011; Lei et al. 2013).
- **Very rapid envelope loss in massive donors.** Very massive stars located within the region of the S Doradus instability strip or approaching the Humphreys–Davidson limit could be subject to strong winds, up to $10^{-4}\ M_\odot\ {\rm yr}^{-1}$, with some stars also having separate stronger mass ejection episodes (e.g., P Cygni or η Carinae). The wild nature of P Cygni-type outbursts in single stars is not yet fully understood. The rapid mass loss observed in stars close to the S Doradus instability strip may be triggered by opacity jumps (Jiang et al. 2018). Similar rapid envelope loss has been argued to be either triggered or enhanced by the presence of a companion in massive pre-CE donors (Eggleton 2002).
- **Interaction-triggered envelope loss.** In eccentric binaries or in triples, interactions can produce rapid orbital changes that lead to violent collisions between companions. Instead of approaching RLOF gradually, in these cases very rapid mass loss can be initiated, leading to the removal of a significant part of the envelope (Ivanova et al. 2017). This mass loss can be several percent of a stellar mass, taking place on a dynamical timescale of the donor. For example, the great eruption of η Car in the 19th century has been argued to be either the mass loss during periastron passage (Kashi & Soker 2010), or a binary merger, as it could have been previously a triple system (Balick 1999; Portegies Zwart & van den Heuvel 2016).
- **Pulsations, AGB Superwinds.** The Reimers wind mass-loss rate has been argued to be boosted by pulsations, e.g., for donors that are close to the Cepheid instability strip when they fill their Roche lobes (Eggleton 2002). The physical motivation is that a pulsation can supply greater acceleration to the outer layers if the potential well is shallower than for a single star. AGB superwinds in single stars can be as large as $10^{-4}\ M_\odot\ {\rm yr}^{-1}$, and are also

thought to be triggered by pulsations (e.g., McDonald et al. 2018). The AGB superwind may potentially also be enhanced by the presence of a close companion.

- **Rotation-driven wind.** Another potential driving mechanism for enhanced winds is connected to the rotational velocity of the star. Since giant stars are normally expected to be slowly-rotating, rapid rotation requires special circumstances, e.g., if the donor star has been tidally spun-up. The physical motivation is that the wind is enhanced in an equatorial direction due to a lower effective gravity at the equator. One of the most-used approaches is to include the following boost factor (Heger et al. 2000):

$$\dot{M}_{\rm w}(\Omega_{\rm rot}) = \dot{M}_{\rm w}(\Omega_{\rm rot} = 0) \times \left(\frac{1}{1 - v/v_{\rm crit}}\right)^{\xi}, \ \xi \approx 0.43, \tag{5.13}$$

where v is the equatorial surface rotation rate and $v_{\rm crit}$ is the critical rotation rate (see Heger et al. 2000, for more details). It is important to realize that the ratio $v/v_{\rm crit}$ does not exceed 0.6 in a tidally locked binary for donors below 20 $M_\odot$ and, for mass ratios below 10, may maximally approach 0.7 for donors up to 100 $M_\odot$ (Marchant et al. 2017). For all mass ratios below 100, the ratio $v/v_{\rm crit} < 0.72$, which follows from the Roche-lobe radius approximation by Eggleton (1983). Hence, even though a boost is expected, it does not exceed a 73% increase.

Mass loss from the donor without much capture by the companion causes the binary to widen. Indeed, for the case when mass is lost as a fast, isotropic wind with the specific angular momentum of the donor, $\gamma = M_{\rm comp}/M_{\rm d}$, and with the mass-loss rate $\dot{M}_{\rm w} = -\dot{M}_{\rm d}$, the orbital separation evolution from Equation (5.11) simplifies to

$$\frac{\dot{a}}{a} = \frac{\dot{M}_{\rm w}}{M_{\rm d} + M_{\rm comp}}. \tag{5.14}$$

Enhanced mass loss before RLOF may not only change the envelope mass of the donor, and hence increase the stability of RLOF against CEE, but may even lead to avoidance of RLOF entirely.

5.6 Roche-lobe Overflow and L_1 Mass Transfer

Roche-lobe overflow mass transfer can be stable or unstable depending on the response of the donor to mass loss with respect to that of its Roche lobe. This response can be determined using stellar evolution codes, but a simpler way to evaluate the stability of RLOF is to compare the change in the donor's radius to the change in its Roche-lobe radius with mass loss. This provides a useful analytical framework for understanding stability criteria, though as we will see, misapplication of it can produce incorrect results.

The following form of this analysis was first introduced by Webbink (1985; see also for more details Hjellming & Webbink 1987). The response of the donor radius to the mass loss is written as a power law, $R_d \propto M^{\zeta_d}$. Due to the mass transfer, the Roche-lobe radius changes as well, and its response to the mass loss is written as $R_L \propto M^{\zeta_L}$. ζ_d and ζ_L are known as mass–radius exponents:

$$\zeta_d \equiv \frac{\partial \log R_d}{\partial \log M}, \quad \zeta_L \equiv \frac{d \log R_L}{d \log M}. \tag{5.15}$$

ζ_d describes how the donor would respond to mass change if it were not constrained by a Roche lobe. Multiple types of the ζ_d exponent are used depending on the timescale of the mass loss, since the physical response of the donor depends on this timescale. On a dynamical timescale it is common to assume that the star responds *locally* adiabatically to mass loss, i.e., with a fixed specific entropy profile. The validity of this approximation will be discussed further below.

To find ζ_L, one can use the best knowledge available to find how the orbital separation (and accordingly the Roche lobe of the donor) would change upon mass loss. The change in orbital separation, and accordingly the Roche-lobe radius, depends on how much mass and angular momentum are lost. Depending on the adopted mode of mass transfer, donor material could either be moved from one star to another, or it could be lost from the system, either from the vicinity of the donor or the companion. Angular momentum then could also be carried away from the binary system with the lost material, or it could be lost due to magnetic braking or gravitational-wave radiation. It could also be redistributed (e.g., transferred to a companion, or a circumbinary disk). In the simplest case of fully-conservative mass transfer—i.e., neither mass nor angular momentum are lost from the binary—the Roche-lobe response is (Tout et al. 1997):

$$\zeta_L = 2.13q - 1.67. \tag{5.16}$$

The donor's radius response can be obtained for various types of donors—with a radiative or an outer convective zone, hydrogen or helium-rich envelope, and so on (for polytropes or composite polytropes, see Hjellming & Webbink (1987); Soberman et al. (1997); for He stars, see Belczynski et al. (2008)). The two main types of response are the so-called "radiative" and "convective" responses. For analyses of the dynamical stability, the responses are usually found using a so-called "adiabatic" approximation (Hjellming & Webbink 1987; Ge et al. 2010, 2015). Note that here the term "adiabatic" refers to individual mass elements rather than the star as a whole. Since the envelope layers are assumed to have no time to exchange energy, their specific entropies are fixed, and so the response due to the mass loss can be computed by using the layers' initial entropies. Roughly speaking, donors with radiative outer zones can be said to have positive $\zeta_{d,ad}$, and donors with convective outer zones have $\zeta_{d,ad} < 0$. Note that a positive ζ_d means that the donor shrinks upon mass loss.

The mass transfer is dynamically unstable if, upon rapid mass loss, the degree by which the donor radius exceeds its Roche lobe increases:

$$\zeta_{d,ad} < \zeta_L \text{ for instability.} \tag{5.17}$$

For each type of donor (and its adiabatic response) and for each type of mass transfer (and how conservative it is), a critical mass ratio $q_{\rm crit}$ can be found, such that if the mass ratio in the considered binary satisfies $q > q_{\rm crit}$, the mass transfer is dynamically unstable.

The most common type of donor in common-envelope cases is a hydrogen-rich donor with a deep convective envelope. The adiabatic response predicted by polytropic models of such stars depends on the fraction of the stellar mass in the convective envelope (see, e.g., Hjellming & Webbink 1987, in which the core has gravitational mass but is otherwise inert, and the envelope is a $\gamma = 5/3$ polytrope). If the model is fully convective then $\zeta_{\rm d,ad} = -1/3$, i.e., the star expands upon mass loss. If the fraction of the stellar mass in the core is greater than approximately 0.21, then the predicted adiabatic response from a condensed polytrope model is that the star will contract upon mass loss (Hjellming & Webbink 1987). If we adopt $\zeta_{\rm d,ad} = 0$, and assume fully conservative mass transfer as in Equation (5.16), we find $q_{\rm crit} \approx 0.8$. In the past, this automatically led to the conclusion that any mass transfer produced when the primary (initially more massive star) overfills its Roche lobe causes CEE to take place. Specifically, in the past this led to the proposal of scenarios to help explain how observed close double white dwarf systems could be formed through two successive phases of unstable mass transfer (e.g., Nelemans et al. 2000). However, for the condensed polytropic models of Hjellming & Webbink (1987), and assuming conservative mass transfer, $q_{\rm crit} > 1$ when the core mass exceeds approximately half the mass of the star (e.g., for low-mass AGB stars). Moreover, non-conservative mass transfer tends to make RLOF more stable (see, e.g., Podsiadlowski et al. 2002; Soberman et al. 1997; Kalogera & Webbink 1996), unless more specific angular momentum is carried away by the mass lost than is typically assumed.

For donors with an outer radiative zone, the critical mass ratio is typically found using detailed binary evolution calculations to be in the range 3.5–4 (see, e.g., Hjellming & Webbink 1987; Podsiadlowski et al. 2002).[1] The initial mass loss rate is well below dynamical, simplifying the calculations, but it grows with time eventually to runaway. This is known as a delayed dynamical instability (e.g., Hjellming & Webbink 1987; Ge et al. 2010). Hence it has been appreciated for many years that radiative donors could lose a significant amount of mass prior to the instability, as is still thought to be the case. However, for convective donors the growth timescale of the instability was generally thought to be dynamical, i.e., effectively immediate, which is no longer the current understanding.

We have indicated above that much of the community has, broadly speaking, often expected RLOF to be more unstable than predicted even by approximate adiabatic models (see also, e.g., the discussions in Podsiadlowski 2001; Han et al. 2002). Moreover, some of the caveats noted by the developers of approximate mass-transfer stability models have not been as widely appreciated as they should have

[1] The Hjellming & Webbink (1987) paper is known for its study of mass-transfer stability via polytropic models, but the delayed dynamical instability case they present is based on a full binary evolution calculation from Webbink (1977). In that case the mass ratio leading to a delayed dynamical instability is 3.

been. For example, Osaki (1970) stated that assuming adiabatic response for rapid mass transfer may well be a qualitatively misleading oversimplification, because the surface layers have sufficiently short thermal timescales to respond on the timescale of the mass loss. Also, Paczyński & Sienkiewicz (1972) showed how predictions from their model for dynamical timescale mass transfer could be sensitive to changes in the thermal treatment of the convective envelope. Perhaps confusion in works citing these models has been exacerbated by the ambiguity inherent when theorists refer to "the thermal timescale" of a star. Real stars respond on multiple thermal timescales, and the outer layers of a donor star can thermally adjust much more rapidly than the whole-star thermal timescale (see, e.g., Podsiadlowski et al. 2002; Woods et al. 2012).

Because real stars do not respond in a truly adiabatic fashion, mass transfer calculations using one-dimensional stellar evolution codes find that RLOF is significantly more stable than predicted when using the locally adiabatic approximation (see, e.g., Woods et al. 2012). For example, all the binary-evolution calculations for mass transfer from low-mass giant donors in Han et al. (2002) yield $q_{\rm crit} > 1$, and the stability threshold reached as high as $q_{\rm crit} \approx 1.3$ (see also Woods et al. 2012). In addition, the $q_{\rm crit}$ values calculated by Han et al. (2002) do not follow a monotonic trend with increasing core-mass fraction, as had been expected from the condensed polytropes in Hjellming & Webbink (1987).

Another effect that contributes to increased mass-transfer stability is that during RLOF not all donor material located outside of the Roche lobe can be immediately transferred through L_1, as the gas streaming motion is limited by the L_1 cross-section; see Figure 5.1. Thus it is possible for the donor to expand slightly beyond its Roche lobe before L_1 outflow begins to compensate for the expansion. As a result, the donor remains mildly overflowing, while the mass transfer does not run away.

Using the physical limitation derived by Pavlovskii & Ivanova (2015) on the rate of mass flow through L_1, the critical mass ratios for donors with fully developed convective envelopes are found to be in the range 1.5–2.2 for most donors. We note that these values are for conservative mass transfer, and that non-conservative mass transfer would lead to larger critical mass ratios (i.e., more stable mass transfer). Pavlovskii & Ivanova (2015) found that, during the development of the outer convective envelope, convective donors are in a transitional regime between the canonical "radiative" and "convective" responses. For donors where the outer convective envelope is non-negligible but is not yet fully developed, the critical mass ratio is just below 3.5, similar to the case of radiative donors.

In case of massive donors up to 80 $M_\odot$, the Pavlovskii & Ivanova (2015) calculations in the L_1 stream-limited regime indicated that the effective critical mass ratio may even be as high as $q_{\rm crit} = 8$. For these cases, as with radiative donors, by the time the instability eventually takes place a significant fraction of the envelope's material has already been removed. This is a period of rapid thermal-timescale mass transfer which can be highly non-conservative. Note that for these very luminous and extended donors, the thermal timescale and dynamical timescale can become similar. In the limit of $\beta = 0$, and for material that is lost with the

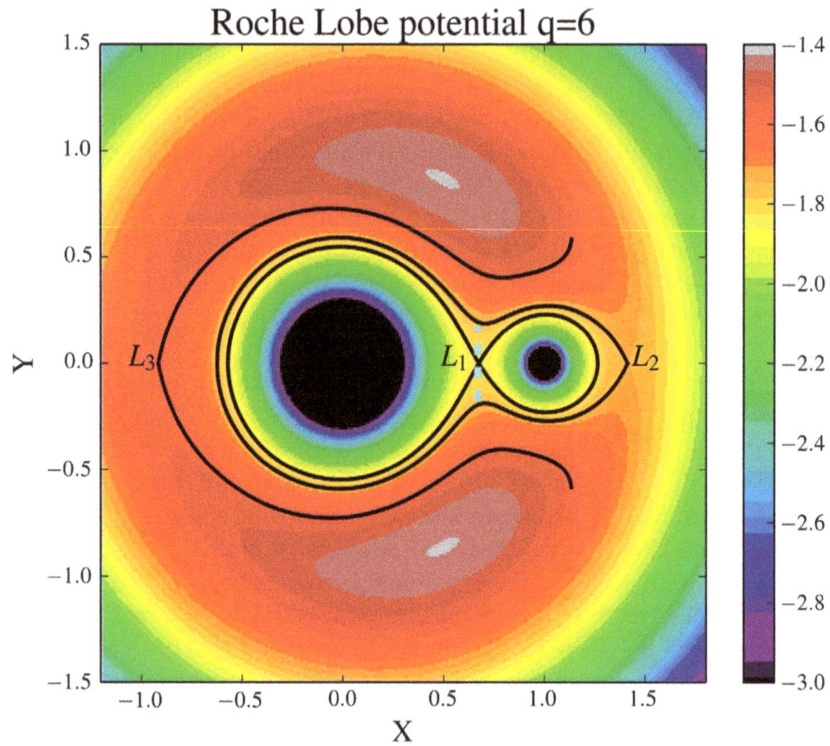

Figure 5.1. Dimensionless effective potential for a binary (the effective potential in the corotating frame, divided by $G(M_1 + M_2)/a$). Shown is the equatorial plane for a binary system with a mass ratio of 6. L_1, L_2, and L_3 are the main saddle points determining the mass transfer; black lines show the equipotentials going through each of those. The L_3 equipotential is only drawn partially, to indicate the extreme shape of the more massive star when it reaches L_3. Values of the effective potential below −3 are capped and shown only in black. The cyan dashed line indicates the cross-section of the stream of material from the donor in the neighborhood of L_1 if the donor fills to the equipotential surface which passes through L_2.

specific angular momentum of the companion, $\gamma \to M_d/M_{comp}$. In this case Equation (5.11) simplifies to

$$\frac{\dot{a}}{a} = 2\dot{M}_d \frac{M_d - M_{comp}}{M_d M_{comp}} = 2\frac{\dot{M}_d}{M_d}(q - 1). \tag{5.18}$$

Hence if a substantial amount of envelope mass has been removed, and the mass ratio q of the donor and the companion becomes less than one, the orbit can start to widen, and the CEE can be avoided.

While the start of dynamical-timescale mass transfer, and hence of a common envelope event, in convective donors is delayed as compared to the initial moment of RLOF, it is important to note that this effect is not the same in terms of driving physics as the well-known delayed dynamical instability mentioned above for radiative donors. For convective donors the delay is driven by competition between mild ongoing RLOF and removal of the material through the neighborhood of L_1, while for radiative donors it is driven by changes in the entropy of the removed material due to the steeply curving entropy profile inside the initial donor.

For all situations in which RLOF becomes unstable after an extended phase of mass transfer, leading to a CEE, this breaks assumptions commonly used in both three-dimensional hydrodynamic simulations of CE events and binary population synthesis models. The structure of the donor star at the onset of RLOF, before it is perturbed by prolonged mass transfer, is generally currently used to provide the initial condition for a CE phase, whether for the energy formalism or for three-dimensional simulations. Future calculations which take into account the delayed onset of a common envelope event may lead to outcomes different from those performed at present.

To summarize these results regarding the stability of mass transfer: (a) a smaller fraction of mass-transferring binary systems is expected to enter a CE phase compared to works which adopt simplified assumptions, since instability from L_1 mass transfer requires more extreme mass ratios than predicted when using those simplified assumptions; (b) a significant amount of envelope mass can be removed during stable L_1 mass transfer before L_2 outflow develops; (c) both convective and radiative donors can experience a delay between the initiation of RLOF and the start of a CEE, and the effect of this delay on the donor's structure has not generally been considered in the energy formalism or in simulations.

5.7 Mass Loss via Outer Lagrangian Points

While there is no unique definition of when and why the CEE definitely starts, the start of L_2 or L_3 mass transfer usually indicates that the binary is ready to transition into a CEE. The specific angular momentum that stellar material would have if in a co-rotation with either L_2 or L_3 is substantially larger than that associated with the donor or companion, γ is becoming larger (see Equation (5.8)) and, as follows from Equation (5.11), the rate at which the orbit shrinks speeds up as soon as L_2 or L_3 outflows develop, even if the total mass loss rate is the same as before.

For L_3 outflows in cases when L_3 is on the side of the mass-losing donor, it can be shown geometrically that the cross-section of the donor in the neighborhood of L_1 becomes very large, a substantial fraction of the donor's radius; see the example in Figure 5.1. Stellar evolution calculations performed through RLOF that take into account the stream's density at L_1 and the stream's cross-section indicate that, by the time the donor has expanded to L_3, the mass transfer rate via L_1 is so high that it is becoming a few percent of the donor's dynamical mass loss rate, $\dot{M}/\tau_{\rm dyn}$ (Pavlovskii & Ivanova 2015).

Once the orbital separation starts to change at the rate of a few percent per orbit, the Roche approximation ceases to be valid, as the Euler term of the fictitious force, which is ignored in the Roche formalism along with the Coriolis term, becomes comparable to the centrifugal term. The rate at which the orbit shrinks becomes comparable to the rate which indicates the start of the plunge-in, as discussed in Chapter 2. Hence expansion of the donor to its outer Lagrangian point seems to be indicative of the start of dynamical mass transfer, and so of the canonical CE plunge-in phase.

Understanding the evolution of a binary with L_2 outflows via the companion's outer Lagrangian points requires three-dimensional calculations, where the mass loss from the donor via L_1 becomes itself dependent on how much angular momentum is being carried away through L_2. We will refer to the case of outflows on the side of the companion as L_2 outflows, even though for a general case with an arbitrary mass ratio these could be formally termed L_3 outflow.

In three-dimensional simulations performed with mass ratios $q > 3$, L_2 outflow very quickly (within ~10–100 orbits) leads to runaway mass transfer and CEE (Nandez et al. 2014; MacLeod et al. 2018). However, not all possible cases have been explored, so we cannot say yet that this result guarantees that any L_2 outflow will necessarily and rapidly lead to the start of a CEE.

A promising way to predict the longer-term role of L_2 mass loss is to compare the angular momentum parameter γ extracted from three-dimensional simulations to the analytically expected values for γ_{L2} found under the assumption that the lost material would have the specific angular momentum of L_2. MacLeod et al. (2018) found that while the material that is lost from the binary carries away more specific angular momentum than the companion would have, it may take away either more or less than the L_2 value during the same simulation, making predictions without simulations very difficult. Chen et al. (2018) has also found quite a wide range of time-averaged values of γ for the outflows as they leave the binary, varying anywhere between 0.8 and 7.8.

The fact that $\gamma > \gamma_{L2}$ at early stages as found by Chen et al. (2018) is presumably the reason why the simulations of (Nandez et al. 2014) showed a faster runaway coalescence for simulations of V1309 Sco than would be predicted by an analytic estimate that uses L_2 specific angular momentum loss. Clearly, a better understanding of how quickly L_2 outflow leads to the start of the plunge-in is needed, as the evolution of the binary system through the L_2 outflow regime may help to interpret properly the rising part of the light curves of luminous red transients (Pejcha 2014; Metzger & Pejcha 2017; MacLeod et al. 2018). For example, L_2 mass loss could be one of the ways in which V1309 Sco started its merger, as the dimming of its light curve prior the outburst can be interpreted as due to excretion disk formation (Tylenda et al. 2011).

Despite the fact that L_2/L_3 outflows likely signal the oncoming instability of the mass transfer, some observed binary systems which are known to have undergone high mass-loss rates for many thousands of years, e.g., SS 433, are argued to be experiencing L_2 mass loss (Bowler 2010; Waisberg et al. 2019).

5.8 The Onset of Double-core Common-envelope

Historically, common-envelope evolution was sometimes initially referred to as double-core evolution. This "double-core" terminology has since taken on a more specific meaning; it now refers to common-envelope phases involving two stars which each possess well-developed cores, and for which a successful envelope ejection would remove both envelopes and expose the stellar cores of both stars (Brown 1995; Dewi et al. 2006). That is, both of the pre-CE stars involved have structures appropriate for the role of a canonical CE donor, and both cores inspiral through the combined envelopes of both stars.

The onset of double-core CE phases is thought to be different from typical CE phases. If both stars in a binary system are to be post-main-sequence, evolved stars with well-developed cores at the same time, then we generally expect the initial masses of the stars to be similar. For such similar component masses, mass transfer is not normally expected to become dynamically unstable as a direct consequence of RLOF (as described in Section 5.6), nor tidal instability (Section 5.2).

An additional scenario that has been proposed as leading to double-core common-envelope evolution involves stable RLOF through the inner Lagrangian point, eventually leading to orbital instability because of mass loss from one or both of the *outer* Lagrangian points (cf Section 5.7). For the special, rare, case of identical initial stellar masses, one could imagine both stars simultaneously filling their Roche lobes to the inner Lagrangian point, leading to no net mass transfer, and the stars then continuing to expand as they evolve until they overflow their outer Lagrangian points. For non-identical initial masses, sufficiently rapid stable mass transfer from the primary star onto the secondary star can cause the secondary star to expand on a thermal timescale, as the accreting envelope is driven out of gravothermal equilibrium (see, e.g., Pols 1994). That accretion-driven envelope expansion may cause the secondary to overflow its own Roche lobe, and then one or both of the components to overflow their outer Lagrangian points, leading to orbital instability (see, e.g., Dewi et al. 2006).

This type of potential CE onset is particularly difficult to model confidently. As with the delayed dynamical instability (for which see Section 5.6), this type of CE onset involves an extended phase of pre-instability stable RLOF which is unsuited to, and is prohibitively expensive for, purely hydrodynamic 3D simulations. Moreover, 1D stellar-evolution models make simplified assumptions in treating the accretion onto the secondary star; commonly the specific entropy of the accreted matter is assumed to be identical to the specific entropy at the surface boundary condition of the accreting star. This assumption affects the extent to which the accreting secondary star expands. Hence we cannot be confident how many binary systems which might reach a RLOF-driven thermal-timescale contact phase actually do so, and we are even less confident about how many are driven to orbital instability and so into a double-core CE phase or merger. Nonetheless, there are observational indications that double-core CE ejection does sometimes occur (see, e.g., Justham et al. 2011 and Section 10.2.3).

5.9 The Effects of Pre-plunge-in Evolution on CE Evolution

Let us summarize the main points regarding pre-CE evolution:

- Tidal spin-up may provide some rotational energy to the envelope above that which this donor would reach in the course of single-star evolution. Tidal heating may deposit frictional energy and increase the internal energy of the envelope. The energy distribution will not be exactly the same as in the case of a single unperturbed star.

- Donors expanding rapidly due to their evolution may start RLOF out of synchronization.
- Enhanced pre-RLOF mass transfer reduces the donor's envelope mass and may lead to a delayed start of CE. The total angular momentum of the system does not have to be conserved during this stage. The orbit may shrink or widen, depending on how conservative or non-conservative the mass transfer onto the companion is.
- The stability of mass transfer has been underestimated in the past. Increased stability of the mass transfer implies that in the case of a standard RLOF for which the response of the accretor is irrelevant, only binary systems with a mass ratio more than one should enter a CEE, perhaps higher.
- However, even detailed one-dimensional calculations used to determine mass-transfer stability are based on approximations which break down close to instability. For example the Roche approximation, and how the mass transfer is treated.
- The criterion for mass transfer instability at the initial moment of RLOF is not sufficient to guarantee that a binary will enter a CEE. The onset of L_2 equipotential overflow for the donor is a more robust criterion, and starting L_2 outflows requires studies specific to each system.

Whenever the mass of the envelope has been decreased by pre-CE evolution, the binding energy of that envelope decreases too. The values of $\alpha_{\rm CE}$ inferred from observed post-CE binaries are overestimated if enhanced pre-CE mass loss is not taken into account. Neither L_1, L_2, nor L_3 mass loss uses only the orbital energy to remove mass from the system, hence these stages are not easy to self-consistently interpret within the energy formalism.

References

Balick, B. 1999, in ASP Conf. Ser. 179, Eta Carinae at the Millennium, ed. J. A. Morse, R. M. Humphreys, & A. Damineli (San Francisco, CA: ASP), 373

Bear, E., & Soker, N. 2010, NewA, 15, 483

Belczynski, K., Kalogera, V., Rasio, F. A., et al. 2008, ApJS, 174, 223

Bowler, M. G. 2010, A&A, 521, A81

Brown, G. E. 1995, ApJ, 440, 270

Chen, X., Han, Z., & Tout, C. A. 2011, ApJL, 735, L31

Chen, Z., Blackman, E. G., Nordhaus, J., Frank, A., & Carroll-Nellenback, J. 2018, MNRAS, 473, 747

Claret, A. 2004, A&A, 424, 919

Darwin, G. H. 1879, RSPS, 29, 168

Dewi, J. D. M., Podsiadlowski, P., & Sena, A. 2006, MNRAS, 368, 1742

Eggleton, P. P. 1983, ApJ, 268, 368

Eggleton, P. P. 2002, ApJ, 575, 1037

Farihi, J., Becklin, E. E., & Zuckerman, B. 2005, ApJS, 161, 394

Ford, E. B., Kozinsky, B., & Rasio, F. A. 2000, ApJ, 535, 385
Ge, H., Hjellming, M. S., Webbink, R. F., Chen, X., & Han, Z. 2010, ApJ, 717, 724
Ge, H., Webbink, R. F., Chen, X., & Han, Z. 2015, ApJ, 812, 40
Han, Z., Podsiadlowski, P., Maxted, P. F. L., Marsh, T. R., & Ivanova, N. 2002, MNRAS, 336, 449
Harrington, R. S. 1968, AJ, 73, 190
Harrington, R. S. 1969, CeMec, 1, 200
Heger, A., Langer, N., & Woosley, S. E. 2000, ApJ, 528, 368
Hjellming, M. S., & Webbink, R. F. 1987, ApJ, 318, 794
Hut, P. 1980, A&A, 92, 167
Ivanova, N., da Rocha, C. A., Van, K. X., & Nand ez, J. L. A. 2017, ApJL, 843, L30
Jiang, Y.-F., Cantiello, M., Bildsten, L., Quataert, E., Blaes, O., & Stone, J. 2018, Natur, 561, 498
Justham, S., Podsiadlowski, P., & Han, Z. 2011, MNRAS, 410, 984
Kalogera, V., & Webbink, R. F. 1996, ApJ, 458, 301
Kashi, A., & Soker, N. 2010, ApJ, 723, 602
Kozai, Y. 1962, AJ, 67, 591
Kushnir, D., Zaldarriaga, M., Kollmeier, J. A., & Waldman, R. 2017, MNRAS, 467, 2146
Lei, Z.-X., Zhang, F.-H., Ge, H.-W., & Han, Z.-W. 2013, A&A, 554, A130
Lidov, M. L. 1962, P&SS, 9, 719
Mace, G. N., Mann, A. W., Skiff, B. A., et al. 2018, ApJ, 854, 145
MacLeod, M., Ostriker, E. C., & Stone, J. M. 2018, ApJ, 863, 5
Marchant, P., Langer, N., Podsiadlowski, P., et al. 2017, A&A, 604, A55
McDonald, I., De Beck, E., Zijlstra, A. A., & Lagadec, E. 2018, MNRAS, 481, 4984
Metzger, B. D., & Pejcha, O. 2017, MNRAS, 471, 3200
Michaely, E., & Perets, H. B. 2019, MNRAS, 484, 4711
Nandez, J. L. A., Ivanova, N., & Lombardi, J. C. Jr 2014, ApJ, 786, 39
Naoz, S. 2016, ARA&A, 54, 441
Nelemans, G., Verbunt, F., Yungelson, L. R., & Portegies Zwart, S. F. 2000, A&A, 360, 1011
Osaki, Y. 1970, ApJ, 162, 621
Paczyński, B., & Sienkiewicz, R. 1972, AcA, 22, 73
Pavlovskii, K., & Ivanova, N. 2015, MNRAS, 449, 4415
Pejcha, O. 2014, ApJ, 788, 22
Podsiadlowski, P. 2001, in ASP Conf. Ser. 229, Evolution of Binary and Multiple Star Systems, ed. P. Podsiadlowski, S. Rappaport, A. R. King, F. D'Antona, & L. Burderi (San Francisco, CA: ASP), 239
Podsiadlowski, P., Rappaport, S., & Pfahl, E. D. 2002, ApJ, 565, 1107
Pols, O. R. 1994, A&A, 290, 119
Portegies Zwart, S. F., & van den Heuvel, E. P. J. 2016, MNRAS, 456, 3401
Reimers, D. 1975, MSRSL, 8, 369
Soberman, G. E., Phinney, E. S., & van den Heuvel, E. P. J. 1997, A&A, 327, 620
Tout, C. A., Aarseth, S. J., Pols, O. R., & Eggleton, P. P. 1997, MNRAS, 291, 732
Tout, C. A., & Eggleton, P. P. 1988, MNRAS, 231, 823
Tylenda, R., Hajduk, M., Kamiński, T., et al. 2011, A&A, 528, A114
Waisberg, I., Dexter, J., Petrucci, P.-O., Dubus, G., & Perraut, K. 2019, A&A, 623, A47
Webbink, R. F. 1977, ApJ, 211, 486
Webbink, R. F. 1985, Stellar Evolution and Binaries (Cambridge: Cambridge Univ. Press), 39

Woods, T. E., Ivanova, N., van der Sluys, M. V., & Chaichenets, S. 2012, ApJ, 744, 12
Zahn, J. P. 1975, A&A, 41, 329
Zahn, J.-P. 1977, A&A, 57, 383
Zahn, J.-P. 1989, A&A, 220, 112
Zahn, J. P. 2008, in EAS Publications Series, Vol. 29, ed. M. J. Goupil, & J. P. Zahn (Les Ulis Cedex: EAS Sciences), 67

Natalia Ivanova, Stephen Justham and Paul Ricker

Chapter 6

The Plunge-in

The initial, fast, phase of the inspiral is known as the plunge-in. This is the only phase that occurs on a timescale comparable to the initial dynamical timescale of the envelope. Considering only the available mechanical, gravitational potential, and internal energies of the envelope may often be an adequate approximation during the plunge. For many donor stars the duration of the phase is far shorter than the thermal timescale of the envelope, so the energy budget assumed in the standard parameterization of common-envelope evolution (CEE) may well be reasonable for the plunge. Hence, for those cases, purely hydrodynamic 3D calculations may be able to provide a good description of this phase. Partly because to date this phase has been most studied with 3D hydrodynamical codes, it is sometimes mistaken as being representative of the whole of CEE.

6.1 The Start of the Plunge-in and the Initial Conditions

The plunge-in is the stage when the orbital decay occurs on a dynamical timescale. Examples of the evolution of the separation between the donor core and the companion are shown in Figures 6.1 and 6.2. The latter figure also includes symbols showing how the rate of evolution of the separation compares to the orbital period, i.e., quantifying to what extent the orbital changes are dynamical. The rapid change in separation starts soon after the point at which $|\dot{a}|P_{\rm orb}/a = 0.1$; this may provide a useful quantitative indication for the beginning of the plunge. Here $P_{\rm orb}$ is the the radial period, which is the time necessary to travel from pericenter to apocenter and back. In existing simulations the plunge-in starts quickly, within ten to a hundred orbits, after the initial binary starts to experience outflows through its outer Lagrange points, although this might be not physically realistic (and outflow from the outer Lagrange points may not be correct initiation criterion; see Section 5.7).

It can take hundreds of orbital periods to evolve a system from the time when the donor fills its Roche lobe to the moment when the orbital decay becomes dynamical, even for the relatively prompt case of Darwin-instability driven CEE. This is

doi:10.1088/2514-3433/abb6f0ch6

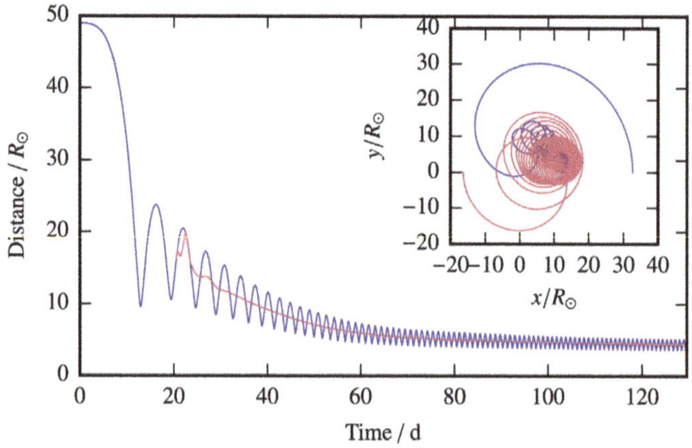

Figure 6.1. Distance between a red giant core and the companion (blue) and semimajor axis (red), in solar radii, over time in days. The inset shows the positions of the red giant core (red) and companion (blue) in the xy plane up to 80 days. The mass of the red giant is $2M_\odot$, its core mass is $0.4M_\odot$, and the companion's mass is $1M_\odot$. At the start of the simulation, the companion was placed on the surface of the red giant at the same y and z coordinates as the red giant core and at a distance of $49R_\odot$ in the x direction. Because of this initial condition, the plunge-in began promptly rather than developing over a number of orbits. This figure is reproduced from Ohlmann et al. (2016).

computationally expensive, and so most three-dimensional simulations start from a much smaller separation. Making that choice introduces artifacts due to the mismatch between the start of the CEE simulation and the expected real initial conditions. Obviously the amount of initial orbital energy and angular momentum in the system is incorrect. The *distribution* of the energy and angular momentum at the moment when the companion starts the plunge will also be wrong. It may be possible to make corrections for those artifacts which are introduced by artificially hastening the onset of the plunge, but we do not know how, and current calculations do not try to do so. It would be an improvement if all such calculations at least estimated the magnitude of the error in total energies that have been introduced by using artificial initial conditions.

When starting from a circular pre-CEE orbit, an approximate way to understand the start of the plunge is that loss of angular momentum from the orbit drives it to higher eccentricity; see the more detailed discussion in Chapter 5. The orbit changes from circular toward a near-radial infall. This apparent eccentricity typically decreases again by the end of the plunge.

However, the trajectory of the companion during the plunge should formally not be described as the Keplerian orbit of two point masses. The mass of the envelope is considerable, and so the mass interior to the orbit of the companion changes significantly during one orbit. Hence terms that are commonly used, such as orbital energy, orbital period, and eccentricity, can be misleading. It is hard to avoid the use of these terms, and we have already used them, but in this Chapter we will aim to rely on more precise definitions for them. For "the orbital energy" we mean the immediate value of the sum of the kinetic potential energies of the core and the

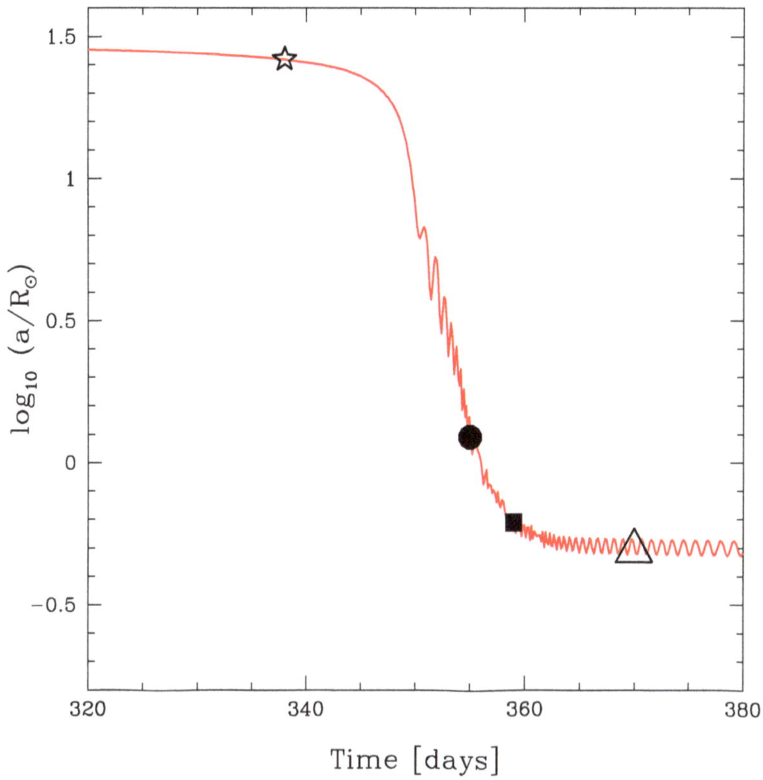

Figure 6.2. Evolution of distances between the donor's core and the companion a in a binary containing a $1.8M_\odot$ red giant donor having a radius of $16.3R_\odot$ with a $0.36M_\odot$ companion. The initial orbital period is 13.86 days, and the initial separation of the binary is such that the donor is just filling its Roche lobe. The apparent non-periodic pattern of the orbital evolution is a consequence of storing only every tenth model. The orbital energy does not oscillate, but due to the nonzero eccentricity, the separation between the giant core and the companion, measured at particular times, shows oscillations. The star symbol shows when $|\dot{a}|P_{\rm orb}/a$ becomes greater than 0.1, indicating the start of the dynamical plunge-in. The circle symbol shows when $|\dot{a}|P_{\rm orb}/a$ again becomes less than 0.1. The square symbol shows when $|\dot{E}_{\rm orb}|P_{\rm orb}/|E_{\rm orb}|$ becomes less than 0.01, which we consider a reasonable indicator for the transition between the plunge and the slow spiral-in. The triangle symbol indicates when $|\dot{E}_{\rm orb}|P_{\rm orb}/|E_{\rm orb}|$ becomes less than 10^{-4}. This figure demonstrates well the start and the end of the plunge-in phase given the above criteria, but it does not show well the oscillations in the orbital separation during the plunge (due to limited sampling). The plot uses the simulation published in Ivanova & Nandez (2016).

companion; the value of this energy oscillates because of the non-Keplerian potential. The "orbital period" refers to the time interval between two successive maxima in the separation between the core and companion. The eccentricity has no formal meaning, since the orbit is not a conic section, but it is loosely interpreted to increase at the start of the plunge, as we write above. Without adopting that terminology we could say that during each pseudo-orbital cycle the difference between the closest approach and largest separation between the core and the companion increases early in the plunge (see Figure 6.1).

6.2 The Plunge Itself: Overview of Three-dimensional Numerical Results

Three-dimensional hydrodynamical simulations of the full volume of the common envelope have been a primary source of insight into the physics of the plunging phase of CEE. The numerical methods and codes used to construct these simulations are discussed in Chapter 4; here we focus on what we have learned from these calculations.

Most 3D simulations have begun with initial conditions corresponding to a situation just before the start of the plunge-in, though the stellar models used have generally not been modified by the mass transfer discussed in Chapter 5. Thus while some simulations have used relaxed initial conditions, even these have generally not been fully consistent with the orbital or mass transfer evolution expected to lead up to the plunge. A recent simulation by Reichardt et al. (2019) does follow many orbits leading up to Roche-lobe overflow and demonstrates the large-scale behavior of the plunge (Figure 6.3).

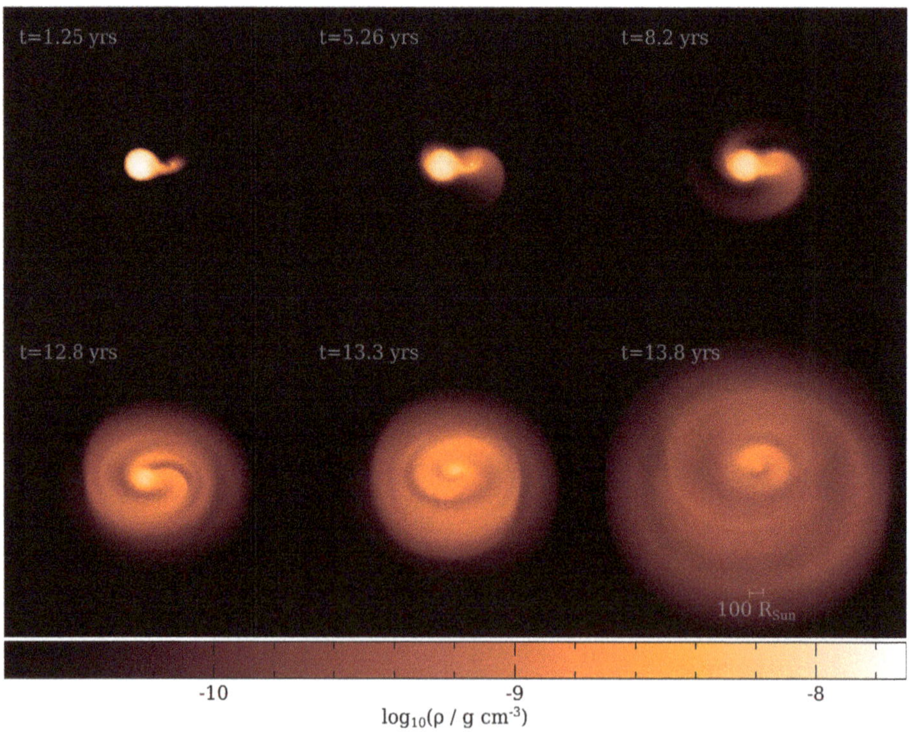

Figure 6.3. Volumetric rendering of a 1.1 million particle SPH simulation for a binary consisting of a $0.88 M_{\odot}$, $90 R_{\odot}$ red giant branch star and a $0.6 M_{\odot}$ compact companion. The frames are 5 au per side. From left to right, top to bottom, each row is a snapshot taken at 1.25, 5.26, 8.2, 12.8, 13.3, and 13.8 yr. This is intended to display the overall evolution of the simulation. The rendering uses an arbitrarily chosen opacity $\kappa = 1.9 \times 10^{-9}$ cm ^{2} g^{-1}. This figure is reproduced from Reichardt et al. (2019). © 2019 The Authors.

Early 3D simulations (Terman et al. 1994; Rasio 1996; Sandquist et al. 1998), reviewed by Taam & Sandquist (2000); established a result that has persisted despite improvements in resolution and numerical methods, namely that in the absence of additional sources of energy like recombination, orbital energy release during the dynamical plunge is not sufficient to eject the full envelope. For example, Terman et al. (1994) used SPH to simulate the evolution of a binary consisting of a $4.97M_{\odot}$ red giant and a $0.94M_{\odot}$ dwarf, finding that only 13% of the envelope was ejected, despite inspiraling by more than a factor of 20. Rasio (1996) also used SPH to simulate the interaction of a $4M_{\odot}$ red giant with a $0.7M_{\odot}$ companion, ejecting only 10% of the envelope within the first 100 orbits. Sandquist et al. (1998) used a nested grid version of the Burkert & Bodenheimer (1993) code to study AGB donors with mass ratios closer to 1:10, finding that between 23% and 31% of the envelope mass was unbound.

Although these early 3D simulations did not eject the full envelope, in each case mass unbinding continued despite the fact that the orbital inspiral timescale had become very long. On the basis of this observation the authors concluded that full envelope unbinding would be completed within ~10–100 yr. However, this extrapolation rested on the (perhaps dubious) assumption that the rate of unbinding would remain at the observed final rate, despite the fact that the orbital inspiral timescale was still rapidly increasing. In each case the resolution was low by modern standards, and the numerical methods employed did not capture shocks well, but both Rasio (1996) and Sandquist et al. (1998) were able to observe spiral shocks, identifying them with the removal of angular momentum from the binary cores and the spinning up of the envelope. Virtually all of the angular momentum transfer occurred during the 1–2 orbits of the initial plunge, while energy transfer from the orbit to the envelope continued at a declining rate over many orbits. As Ivanova et al. (2013) show, the early loss of angular momentum compared to energy is a simple consequence of the fact that, in a Keplerian orbit of radius a, angular momentum $J \propto a^{1/2}$ while energy $E \propto a^{-1}$, so $|dE/dJ| \propto a^{-3/2}$—in other words, as inspiral progresses the amount of energy released per unit of angular momentum increases.

Each of these early papers also demonstrated that the initial dynamical plunge (in which the orbital separation drops by a factor of two or more) lasts about one initial orbital period, despite using different types of initial setup (relaxed versus nonrelaxed, synchronous versus nonsynchronous). Interestingly, simulations with massive donors by Terman et al. (1995) showed different behavior, with the dynamical plunge requiring several initial orbital periods. They also argued that a steep density gradient was needed in order to terminate the inspiral, though they would not have been able to rule out a continued self-regulated inspiral for these cases.

For almost a decade after the Taam & Sandquist (2000) review, little 3D hydrodynamical simulation work was published on common envelopes until Ricker & Taam (2008); who used an AMR simulation with a factor of three higher spatial resolution and a factor of ten larger spatial dynamic range to study the early evolution of a $1.05M_{\odot}$ red giant with a $0.6M_{\odot}$ companion. Based on this simulation and its continuation, Ricker & Taam (2008, 2012) demonstrated the important role of oblique, spiral shocks in transporting angular momentum throughout the envelope during the initial plunge-in and showed that hydrodynamical drag effects were negligible compared to tidal drag.[1] They also demonstrated that accretion onto compact companions during inspiral could be significant. Although they followed a limited number of orbits after the initial plunge, they showed that even with enhanced resolution the full envelope is not ejected during this phase. Together with Passy et al. (2012); who studied a similar donor with different companion masses using SPH and unigrid Enzo simulations, this work led to renewed interest in the problem and began the "modern era" of common-envelope simulations.

Despite improved numerical methods and resolution, these new simulations also failed to eject the full envelope during the initial plunge. For example, Ohlmann et al. (2016) used a moving-mesh code for the first time to study the common-envelope evolution of a $2M_\odot$ giant with a $1M_\odot$ companion. Using a maximum resolution an order of magnitude better than the grid simulations of Ricker & Taam (2012) and Passy et al. (2012); they still found that only 8% of the envelope was ejected after 120 days of evolution. Ohlmann et al. (2016) did observe conditions consistent with the development of shear-driven convection, suggesting the onset of turbulent viscous dissipation like that in the models of Meyer & Meyer-Hofmeister (1979); but they were not able to follow the slow spiral-in to confirm that this would lead to complete envelope ejection.

Nandez et al. (2015) were the first to explore new physics beyond the γ-law hydrodynamics used in previous work. They found that replacing the γ-law equation of state with a tabulated EOS made the initial energy profile of the donor match better the initial profile of the 1D stellar model and allowed for more envelope ejection during the plunge-in phase. Significantly, this allowed them for the first time to include the energy of hydrogen and helium recombination, although without radiation transport they could not treat it in a fully self-consistent manner. Nandez et al. (2015) were able to show that, with this contribution to the internal energy, complete ejection of material above the $0.31M_\odot$ degenerate helium core could be achieved for red-giant donors between 1.0 and $1.8M_\odot$ with a $0.36M_\odot$ companion. This result confirmed that, at least for some masses and types of system, neglect of recombination energy explains why 3D simulations have been unable to achieve envelope ejection.

Another clear result from 3D simulations is that gravitational drag dominates over hydrodynamical drag during the plunge-in phase. Ricker & Taam (2008, 2012) measured the hydrodynamical drag by computing the momentum flux through small control surfaces centered on the companion and donor core. They showed that the net gravitational force due to the envelope exceeds the hydrodynamical drag typically by at least an order of magnitude. They also showed that the gravitational drag (as well as the rate of mass accretion onto the companion) could be as little as 1% of the Bondi–Hoyle–Lyttleton value, interpreting this result as being due to the

[1] It should be noted that the contribution of numerical viscosity to the drag has neither been estimated in any CE simulation, nor has it been compared to values expected from turbulent convection in stars, though in this work the observed inspiral was consistent with the measured gravitational drag.

facts that the companion's motion with respect to the envelope is at best weakly subsonic and that the gas flow near the companion is highly asymmetric, with gas being rapidly flung outward rather than trailing the companion and exerting drag on it (see Figure 3.2). This "slingshot" mechanism transfers angular momentum from the cores to the envelope, and the spiral shocks that result serve to distribute the transferred angular momentum to the rest of the envelope (see Figure 9.2).

Staff et al. (2016) also measured hydrodynamical and gravitational drag for CE simulations involving a red-giant or AGB star and a planet, though they used a ram pressure expression instead of computing a momentum flux for the hydrodynamical drag, and for the gravitational drag they used a modified gravitational capture radius expression. They found that at late times for the red-giant case the gravitational drag declined enough to become comparable to the hydrodynamical drag (though this occurred when the orbital separation was close to the SPH smoothing length). Thus conclusions about the relative importance of gravitational and hydrodynamical effects should be regarded as potentially mass- and mass ratio-dependent.

Using 3D wind-tunnel experiments, MacLeod & Ramirez-Ruiz (2015) and MacLeod et al. (2017) shed further light on the slingshot mechanism, developing a local model for Bondi–Hoyle–Lyttleton accretion from gas with a transverse density gradient that can be used in 1D simulations (see Section 7.2.4). This model has been tested directly by Chamandy et al. (2019); who show that during the initial plunge it accurately predicts the measured drag in a global CE simulation with a 1:8 mass ratio. After the initial plunge, as the orbit circularizes, the companion begins to encounter material it has already disturbed, and the MacLeod et al. (2017) model begins to significantly overpredict the drag force. However, Chamandy et al. (2019) do not compare the model to simulations with smaller mass ratios. In fact, as their Figure 6 demonstrates for a mass ratio of 1:2, initially the gravitational force due to the entire envelope is significantly unbalanced on large scales, suggesting that a local model would not work for this case. As the orbit circularizes and the inspiral slows, the long-range gravitational forces due to the envelope becomes more symmetrical, though again the wind-tunnel result would not be expected to be accurate then because the companion has begun to move into material it has disturbed. Kramer et al. (2020) also compare 3D global simulations to the wind-tunnel model and find good agreement, but they consider very small companion masses and hence large mass ratios.

We conclude from these results that for large donor/companion mass ratios the wind-tunnel model of MacLeod et al. (2017) correctly describes the transfer of energy and angular momentum from the companion's orbit to the envelope during the initial plunge for ideal-gas hydrodynamics. A more general local model, still to be developed, that accounts for the companion's disturbance of the envelope might explain the subsequent evolution, as well as termination of the inspiral. However, for more similar masses, the initial tidal deformation of the envelope is significant, and as a result the gravitational drag on the companion is more global, with significant tidal forces on the companion originating outside the range of a local wind-tunnel simulation. Under these circumstances the companion's disturbance of the envelope is more than a local perturbation, and an accurate model for the drag probably requires an analysis in terms of the tidal modes of the entire envelope, along the lines of the analysis carried out by MacLeod et al. (2019).

Since 3D simulations have rarely produced complete envelope ejection, they have not yielded much insight into why some systems eject the envelope and remain binaries and others merge. However, they do quickly pass through the rapid plunge, and they appear to reach a state in which the inspiral timescale becomes very long. The end of the rapid plunge appears to be related to the tidal drag (see for more details Sections 6.3 and 7.2.4) as well as to spinning up of the envelope (both bound and unbound) to the angular momentum of the initial orbit. The transition to a slow spiral-in appears to be governed for more massive companions by the evacuation of material from the companion's vicinity and the switch to a more local source of drag due to the symmetrization of the envelope, though evacuation does not occur for very low-mass companions (Iaconi et al. 2018). This strongly suggests that the transition is connected to the ratio of envelope to companion mass and the pre-CE envelope structure of the donor. In this light, the hint from Terman et al. (1995) and Ricker et al. (2019) that for supergiant donors there is no rapid plunge would appear to indicate that angular momentum and energy transfer is inefficient when the envelope is very diffuse and the thermal timescale is comparable to the dynamical timescale. In cases for which the thermal timescale is comparable to the dynamical timescale, we might sensibly worry that neglect of radiative transport and losses is troublesome (see Section 4.1.6 and Chapter 7). On the other hand, Hwang & Lombardi (2015) do observe a rapid plunge in massive "twin" CE simulations with individual star masses between 8 and $20 M_{\odot}$, so mass ratio and companion type may also play a role in determining whether there is a plunge and when it ends.

Several mass ejection episodes can take place during the plunge-in. They are different in their nature and will be discussed in more detail in Chapter 8.

6.3 The End of the Plunge-in Phase

If CEE were purely a dynamical-timescale phenomenon, the envelope would be ejected by the end of the plunge-in phase. If that were to happen, it would leave behind a clean binary consisting of the core of the donor and the companion. However, the overwhelming majority of three-dimensional simulations so far have not successfully ejected the envelope. Instead they see a decrease in the rate of orbital decay, leading to a post-plunge phase called the "slow spiral-in" with a relatively low-eccentricity orbit which is the subject of the next Chapter (Chapter 7).

There is no definite criterion for the transition between these phases, but the difference is not merely arbitrary nomenclature. Physically, we want to describe the point at which the approximation of a purely hydrodynamic, energetically closed system is no longer justified. That is, we want to know when the evolutionary timescale becomes longer than dynamical, and certainly before it becomes comparable to the thermal timescale of the expanded envelope.

The physical reason for the transition from plunge-in to slow spiral-in is not fully understood. So far this question has not been addressed carefully in the literature.

Nonetheless, the slowing in the rate of inspiral is presumably related to the transition from global to local tidal dissipation, as a consequence of the expansion of the envelope during the plunge. This expansion decreases the matter density in the region surrounding the new orbit, which decreases the rate at which energy is lost from the orbit through drag (Section 7.2.4). However, the dominant reason for the slowing of the inspiral is typically a more subtle one. The shrinking of the binary, combined with the expansion of the envelope, decreases the strength of the quadrupole interaction between the rotation of the binary and the envelope. In turn the tidal drag on the binary orbit significantly decreases. For further discussion see Chapter 7.

Despite the fact that we do not have a rigorous understanding of the physical reason for the transition, defining quantitative criteria is important for comparing the results of different simulations. We can suggest the following empirical definition. Let us identify the end of the plunge-in with the moment when the amount of mechanical energy transferred from the system formed by the core and the companion to the envelope, during one revolution, drops below some small threshold. Let us introduce the orbital shrinkage timescale as:

$$\tau_{\rm shrink} = |E_{\rm orb}/\dot{E}_{\rm orb}|. \tag{6.1}$$

When comparing simulations, it is better to consider the time when $\tau_{\rm shrink}/P_{\rm orb}$ exceeds some threshold, say 100, rather than rely on qualitative descriptors such as "end of the plunge-in." Typically, as this ratio becomes large, the cost of running 3D hydrodynamical codes eventually becomes prohibitive. Thus, the end of a simulation is more often driven by cost concerns than by physical criteria.

After the end of the plunge-in, the time-averaged separation between the core and the companion continues to slowly shrink. Note that the dynamical timescale of the expanded envelope can be far longer than the orbital timescale of the binary, so the fact that the orbital decay is no longer dynamical is not a guarantee that the decay timescale is longer than the dynamical timescale of the *envelope*. Nonetheless, the timescale of orbital decay after the plunge typically becomes similar to or longer than the dynamical timescale of the expanded common envelope, and can approach or exceed the decreased thermal timescale of the expanded common envelope.

Problems arise for simulations due to this change in relative timescales. First, it is no longer a reasonable approximation to ignore radiative energy losses, or additional energy sources and sinks. Internal energy transport may well also become important. Hence purely hydrodynamic simulations which do not include radiative energy transfer or losses cannot simulate phases which last longer than the thermal timescale of the envelope. Such simulations should also be careful to check whether they have enough resolution to capture convective energy transport.

Second, at this rate of orbital decay, it would take tens of thousands of orbital revolutions for sufficient orbital energy to be transferred to the envelope for it to potentially become unbound. This makes any 3D numerical simulation painfully expensive. More fundamentally, such long simulations suffer from accumulation of errors in energy conservation (see Section 4.3.1). When the accumulated error in the orbital energy exceeds some fraction (say 10%) of the true change in orbital energy from the end of the plunge-in to the end of the CEE, the calculations are no longer quantitatively worthwhile. Of course, since we lack either a clear definition of the end of the plunge-in or the end of the CEE, as well as the change in energy between the two times, for all practical purposes the pain threshold is the only rational criterion for ending a simulation.

The CE evolution now evolves either into a self-regulated slow spiral-in (see Chapter 7), or the envelope is removed through some means. For example, one mechanism for accomplishing this removal is recombination outflows (Section 8.3).

6.4 Plunge-in and 1D Considerations

Three-dimensional hydrodynamics calculations are important but extremely time-consuming, and currently they do not include enough physics to self-consistently model long-timescale phases. These calculations are usually restricted to the simulations of the plunge-in, rarely going far beyond the transition to the self-regulated spiral-in. There is a need for both a larger parameter space of CE calculations, and for modeling of the self-regulated spiral-in. For both purposes, 1D stellar codes have been used. However, this requires a recipe for how to adequately model the plunge-in, an intrinsically three-dimensional phase, using a one-dimensional stellar evolutionary code. We will consider 1D modeling relevant to the slow spiral-in stage in Chapter 7, while here we pay specific attention to attempts to model the plunge-in.

Many barriers must be overcome in the quest to mimic “3D reality” in 1D. Most obvious are:

1. obtaining the rate of orbital energy release and the time-dependent orbital evolution—solving for $\dot{a}(t)$;
2. determining how and where energy should be deposited in the envelope;
3. introducing the binary gravitational potential into the 1D structure of the donor star;
4. matching the solution of a 1D spherically symmetric hydrodynamic problem to the spherically averaged solution of the 3D hydrodynamic problem.

Existing 1D simulations of the plunge-in tackle one or more of these barriers in various combinations and with different degrees of sophistication. The bottom line is that we know of no recipes that allow the plunge-in phase to be realistically modeled using a one-dimensional code. The consequences of the necessary spherical averaging are particularly troublesome for the plunge-in phase. Here we discuss potential problems for each of the barriers that need to be resolved for 1D models of the plunge-in.

6.4.1 Orbital Energy and Orbital Separation

The standard way in 1D to describe the orbital energy, and so to find how much is transferred from the orbit to the envelope, is to use the well-known relation between the orbital separation and the orbital energy

$$E_{\mathrm{orb}} = -\frac{Gm(a)M_{\mathrm{comp}}}{2a}. \tag{6.2}$$

Here $m(a)$ is the mass of the donor within the distance a between the companion and the center of the donor. However, this equation is not directly applicable during the plunge: the enclosed mass $m(a)$ changes during one revolution, it is not located at a point, and the orbit is not closed; hence the orbit cannot be considered using a simplified Keplerian form to provide a simple relationship between da and dE. After the plunge, when most of the envelope is beyond the companion's orbit, the orbital separation and the energy can again be considered to be related in a Keplerian way. However, to link the rate of energy release from the orbit with the rate of change of the orbital separation during the plunge-in, one needs both averaging over one revolution and an educated guess for the eccentricity evolution (for example, considering orbital separation as an osculating parameter; Holgado & Ricker 2019).

The discussion above is related to the *amount* of orbital energy released as the orbit shrinks, or dE_{orb}/da. Assuming we can determine an expression for dE_{orb}/da, the next important question is how *rapidly* that orbital energy is released ($\dot{E}_{\mathrm{orb}}(t)$) or, in other words, how quickly the orbit shrinks ($\dot{a}(t)$). This both determines the timescale of the plunge-in, and the rate at which the 1D envelope has to be supplied with energy.

One promising way to obtain $\dot{a}(t)$ is to use the orbital evolution from a local 3D simulation to fit a toy drag-force model. For example, let us consider a drag model based on local "wind tunnel" accretion simulations (MacLeod & Ramirez-Ruiz 2015; MacLeod et al. 2017). In this model, the companion's orbit is treated as circular, with a gradually decreasing orbital separation $a(t)$ which satisfies the equation

$$\frac{da}{dt} = -F_{\mathrm{drag}}(\varepsilon_\rho)v_\infty \frac{da}{dE_{\mathrm{orb}}}. \tag{6.3}$$

Here F_{drag} is the gravitational drag force on the companion due to the envelope. It is specified using a fit to the local tidal force observed in wind-tunnel simulations with varying vertical density gradients. The background density gradient of the donor model is characterized by $\varepsilon_\rho(a)$, which is parameterized by the ratio of gravitational focusing radius R_{a} to the density scale height H_ρ,

$$\varepsilon_\rho \equiv \frac{R_{\mathrm{a}}}{H_\rho} = -\frac{R_{\mathrm{acc}}}{\rho}\frac{d\rho}{dr}, \tag{6.4}$$

where

$$R_{\mathrm{a}} = \frac{2GM_{\mathrm{comp}}}{v_\infty^2} \tag{6.5}$$

is the Hoyle–Littleton accretion radius for a companion of mass M_{comp}, and v_∞ is the relative velocity of the companion with respect to the envelope. Chamandy et al. (2019) compared the local drag force as found from wind-tunnel experiments with results from a global common envelope simulation, finding them to be in good agreement during most of one simulated plunge-in stage.

With the simplifications described above for $\dot{a}(t)$ and for $dE_{\rm orb}/da$, a simple case of 1D orbital evolution with a fixed envelope structure begins with $a = R$, the donor star's radius, and is terminated when the energy dissipation reaches the binding energy of the envelope,

$$\alpha_{\rm CE}[E_{\rm orb}(R) - E_{\rm orb}(a)] \geqslant E_{\rm bind}(a) = \int_{m(a)}^{M} \left(u_{\rm int}(a) - \frac{Gm(a)}{a} \right) dm. \tag{6.6}$$

Here $u_{\rm int}(a)$ is the specific internal energy of the envelope at radius a, and $\alpha_{\rm CE}$ is the energy formalism efficiency discussed in Chapter 3. Alternatively, the calculations may finish when the remaining mass $m(a)$ no longer fills its Roche lobe.

One recent 1D study used a drag force to model orbital decay during the plunge-in for an unperturbed star made with MESA, and showed good agreement for the orbital evolution with the results of 3D simulations using AREPO (Kramer et al. 2020). We note that since they considered very low-mass companions, $\leqslant 0.08 M_{\odot}$, with a donor of $0.77 M_{\odot}$ ($1\ M_{\odot}$ at its zero-age main sequence), the effect of the companion on the donor might be small, so ignoring feedback on the structure of the star may not be an unreasonable approximation for this case.

For calculations that include feedback on the structure of the donor's envelope, the end of the plunge-in may occur when a significant part of the envelope begins to be ejected. We note that published 3D simulations of CEEs all show qualitatively the same result: the plunge itself is a very fast event, lasting no longer than a few initial orbital periods, if not just one (see, for example, Figures 6.1 and 6.2). The "dynamical" rate of the orbital energy injection for 1D stellar codes may become problematic since those codes were neither created nor tested to solve stellar structures on the dynamical timescale of the star. Recent versions of MESA are able to approximately model envelope ejection in supernovae. However this single imposed large energy input is different from the case of self-consistent calculations for a high rate of ongoing energy injection, as in the case of a CEE which does not become immediately unbound.

6.4.2 Energy Deposition—What and Where

How to model in 1D the way energy is deposited into the envelope of the donor remains a major unsolved question. This question is actually several. How much energy is deposited? In what form (internal or kinetic)? How does the deposition lead to envelope ejection? Where is it deposited? Even in 3D some of these issues are unresolved, but all of them become difficult when trying to construct 1D models.

Above we explain that orbital averaging is a problem for 1D treatments of the plunge, since the binary motions are significantly non-circular. This extends to energy deposition, e.g., if the rate of heating is different at apastron from periastron.

The result of energy deposition in the donor envelope during the plunge-in will likely depend on the relationship between the thermal adjustment and dynamical timescales of the envelope, or on the rate of energy input versus how the envelope can react to it. These timescales will also be different at apastron and periastron.

For example, can an energy transport mechanism like convection convey the energy to the surface at the rate at which it is injected? If yes, for example, as can be expected for some massive donors, the plunge-in may be blended with what we usually call the slow spiral-in stage, as we will discuss in Chapter 7. The energy might be thought of as being redistributed quickly over the envelope above the orbit, and the exact location of the energy deposition might not be crucial. If the transport of energy from the place where it is deposited is not as fast as the energy deposition, the energy will be used locally to do work to expand the gas. In this case the location where the energy is added becomes essential. It has been shown that during the initial rapid heating of the envelope of a low-mass donor in 1D simulations, qualitative changes in the outcome can result from adding the same amount of energy but in different locations in the envelope (Ivanova et al. 2015).

One-dimensional codes must find a way to model how the energy from the orbit is transferred to the envelope that mimics how it actually happens. In 3D hydrodynamic simulations, the direct energy deposition during a dynamical plunge does not appear to primarily occur through local heating, since it produces a relatively moderate increase in the envelope's entropy (Ivanova & Nandez 2016). In 1D stellar codes, the traditional and natural way to mimic orbital energy deposition is by adding a heat source in the energy equation (Meyer & Meyer-Hofmeister 1979). In early 1D calculations this produced a substantially larger entropy increase for the same total energy injected into the envelope. One of the consequences is that recombination of higher-entropy envelope material starts at a lower density, requiring the material to be pushed mechanically to larger distances to begin recombining, as well as increasing the chance that the energy released by recombination escapes from the system. 1D calculations that use a heating term hence do not produce the same result as 3D calculations for a given initial configuration. Modern stellar codes like `MESA` include the hydrodynamics equations and could, in principle, inject energy directly as kinetic energy. However, a comparison of outcomes between 1D and 3D calculations in terms of the envelope's entropy has not yet been made.

Further, the deposition of kinetic energy in 3D simulations is intrinsically asymmetric. In particular, angular momentum deposition proceeds partly through nonlocal tidal interactions and is not spherically symmetric. In 3D, equatorial regions can be accelerated to their local Keplerian velocities while polar regions are barely affected. The local amount of spinning up affects the local amount of energy injection. The consequences of averaging this spin-up into 1D have not yet been studied in the literature.

6.4.3 The Gravitational Potential in 1D

In 1D, introducing an additional mass to represent the inspiraling companion produces the same potential as a thin spherical shell. By Newton's shell theorem, this thin spherical shell would not create a potential inside it, and therefore it does not create a net gravitational force on an object inside it. For this situation the effective gravitational potential $\phi_{\rm env1D}(r)$, can be written as

$$\phi_{\rm env1D}(r) = -\left(\frac{Gm(r)}{r} + f_{\rm ins}\frac{GM_{\rm comp}}{r}\right). \tag{6.7}$$

Here $m(r)$ is the local mass coordinate within the star (excluding the companion), r is the radial coordinate, and $f_{\rm ins}$ indicates the influence of the companion: $f_{\rm ins} = 1$ if the companion orbits within r, or $f_{\rm ins} = 0$ if the companion orbits outside r.

This may in turn falsely suggest that the envelope's gravitational potential energy can be found from

$$\Omega_{\rm env,1D} = -\int_{M_{\rm core}}^{M}\left(\frac{Gm}{r} + f_{\rm ins}\frac{GM_{\rm comp}}{r}\right)dm. \tag{6.8}$$

Let us consider the above using the concept of $\Omega_{\rm env-comp}$, as introduced in Section 3.6. This energy is the part of the potential energy of the envelope due to the companion alone. Let us define $\Omega^{\rm outside}_{\rm env-comp}$ to be the component of $\Omega_{\rm env-comp}$ arising from that part of the envelope outside the orbit of the companion. $\Omega^{\rm inside}_{\rm env-comp}$ is then the part due to the envelope material inside the orbit of the companion, such that $\Omega_{\rm env-comp} = \Omega^{\rm outside}_{\rm env-comp} + \Omega^{\rm inside}_{\rm env-comp}$.

If the thin-shell approximation for the potential energy as in Equation (6.8) were valid at any moment of the CEE, then when the companion was outside the envelope—for example at or before RLOF—it would imply that $\Omega^{\rm outside}_{\rm env-comp} = 0$. But the true value at a distance a is roughly approximated by $\Omega^{\rm outside}_{\rm env-comp} = -GM_{\rm env}M_{\rm comp}/a$.

Some improvement is to treat the companion as a thin shell for the envelope mass outside the orbit, and as a point mass for the envelope inside the orbit of the companion:

$$\begin{aligned}\Omega_{\rm tot,1D} = &-\frac{GM_{\rm core}M_{\rm comp}}{a} - \int_{M_{\rm core}}^{M}\frac{Gm}{r}dm \\ &-\frac{Gm_{\rm env}(r<a)M_{\rm comp}}{a} - \int_{m(r>a)}^{M}\frac{GM_{\rm comp}}{r}dm.\end{aligned} \tag{6.9}$$

However, it has been observed that during the plunge-in of a low-mass giant the thin-shell approximation for the envelope mass outside of the companion's orbit (the last term in the above equation) is still invalid. Close to the orbit, within a torus with thickness 10%–20% of the orbital separation, this approximation produces a potential 50% shallower than the 3D potential (Ivanova & Nandez 2016). At about three times the orbital separation, in the outer envelope, the thin-shell approximation produces a stronger (deeper) potential than 3D. One might come to a better agreement between 3D and 1D by using the appropriate 3D-averaged force due to a point mass on each shell, but this approach has not yet been yet attempted.

6.4.4 ··· and More

The plunge-in is a stage during which the envelope of a donor is expected to react on its dynamical timescale. Use of a 1D stellar code that is capable of dealing with expansion on a dynamical timescale is therefore necessary. Some modern stellar codes like, for example, the publicly available code MESA, include the possibility to switch into the "hydrodynamic" regime. One has to be careful about trusting the outcomes. In addition to all the effects related to averaging as described above (energy injection, potential calculations), fluid phenomena exist that can only occur in multi-dimensional simulations, for example, turbulence, Kelvin–Helmholtz instability, oblique shocks, and more. The role of these phenomena during common-envelope events is not yet fully explored in 3D, and it cannot be said with certainty what will be the effect of their neglect in 1D.

We end by noting that recipes for making post-plunge models could be created. These would enable future one-dimensional modeling of a slow spiral-in stage (for more details see Chapter 7). Note that the post-plunge envelope might contain a density inversion, which is not easily treated by one-dimensional stellar codes. Additionally, if the CEE system considered were to end dynamically in nature, it should not be simulated using 1D codes—the envelope would be becoming unbound on its dynamical timescale, with some interior parts becoming unbound prior to the outer parts. The default boundary conditions, as well as the way in which the stellar structure equations are coded, would usually lead to a problem treating such a star; even if the calculations converged it is not clear to what they would converge.

References

Burkert, A., & Bodenheimer, P. 1993, MNRAS, 264, 798

Chamandy, L., Blackman, E. G., Frank, A., et al. 2019, MNRAS, 490, 3727

Holgado, A. M., & Ricker, P. M. 2019, ApJ, 882, 39

Hwang, J., Lombardi, J. C. J., Rasio, F. A., & Kalogera, V. 2015, ApJ, 806, 135

Iaconi, R., De Marco, O., Passy, J.-C., & Staff, J. 2018, MNRAS, 477, 2349

Ivanova, N., Justham, S., Chen, X., et al. 2013, A&ARv, 21, 59

Ivanova, N., Justham, S., & Podsiadlowski, P. 2015, MNRAS, 447, 2181

Ivanova, N., & Nandez, J. L. A. 2016, MNRAS, 462, 362

Kramer, M., Schneider, F. R. N., Ohlmann, S. T., et al. 2020, A&A, 642, A97

MacLeod, M., Antoni, A., Murgia-Berthier, A., Macias, P., & Ramirez-Ruiz, E. 2017, ApJ, 838, 56

MacLeod, M., & Ramirez-Ruiz, E. 2015, ApJ, 803, 41

MacLeod, M., Vick, M., Lai, D., & Stone, J. M. 2019, ApJ, 877, 28

Meyer, F., & Meyer-Hofmeister, E. 1979, A&A, 78, 167

Nandez, J. L. A., Ivanova, N., & Lombardi, J. C. J. 2015, MNRAS, 450, L39

Ohlmann, S. T., Röpke, F. K., Pakmor, R., & Springel, V. 2016, ApJL, 816, L9

Passy, J.-C., De Marco, O., Fryer, C. L., et al. 2012, ApJ, 744, 52

Rasio, F. A., & Livio, M. 1996, ApJ, 471, 366

Reichardt, T. A., De Marco, O., Iaconi, R., Tout, C. A., & Price, D. J. 2019, MNRAS, 484, 631

Ricker, P. M., & Taam, R. E. 2008, ApJ, 672, L41

Ricker, P. M., & Taam, R. E. 2012, ApJ, 746, 74

Ricker, P. M., Timmes, F. X., Taam, R. E., & Webbink, R. F. 2019, in IAU Symp. 346, High Mass X-Ray Binaries: Illuminating the Passage from Massive Binaries to Merging Compact Objects (Cambridge: Cambridge Univ. Press), 449
Sandquist, E. L., Taam, R. E., Chen, X., Bodenheimer, P., & Burkert, A. 1998, ApJ, 500, 909
Staff, J. E., De Marco, O., Wood, P., Galaviz, P., & Passy, J.-C. 2016, MNRAS, 458, 832
Taam, R. E., & Sandquist, E. L. 2000, ARA&A, 38, 113
Terman, J. L., Taam, R. E., & Hernquist, L. 1994, ApJ, 422, 729
Terman, J. L., Taam, R. E., & Hernquist, L. 1995, ApJ, 445, 367

Natalia Ivanova, Stephen Justham and Paul Ricker

Chapter 7

The Slow Spiral-in

As has been seen in many hydrodynamical simulations, and as was foreseen by pioneering one-dimensional studies of common-envelope (CE) events, envelope expansion in one way or another leads to an increase of the spiral-in timescale, from a value comparable to the orbital time $\tau_{\rm dyn,orb}$ to one substantially larger. In this Chapter, we will discuss various ambiguities in how this stage can be identified in general, and specifically during simulations. The orbital shrinkage timescale can be compared to two different timescales: the new dynamical time of the remaining expanded envelope, $\tau_{\rm dyn,CE}$, or its new thermal timescale, $\tau_{\rm th,CE}$. Accordingly, we can use different terms, either "slow spiral-in" or "self-regulated spiral-in," corresponding to the appropriate timescale.

A number of physical processes, in addition to those considered during the plunge, must now be taken into account to make physical sense of the situation. Energy transport in the envelope now plays a role, both for the energy budget and energy redistribution. Release of recombination energy can play a role in envelope removal, various drag forces affect the timescale of further orbital shrinkage, accretion luminosity starts to become important for envelope heating, and more. Common-envelope evolution (CEE) becomes tangled in complicated physical processes, and interactions between those physical processes occur on similar timescales. The outcome, to a large extent, becomes a victim of which specific approximations are adopted.

7.1 How Should We Identify the Slow Spiral-in in Simulations?

In an ideal world, one would define the slow spiral-in as the stage at which a—the separation between the donor's core and the companion—shrinks on a timescale comparable to a some very well determined timescale that controls the orbital decay. However, recall that after the plunge-in (if it occurs) the dynamical and thermal timescales of the expanded, bound envelope are likely to be comparable. Although many calculations assume one or the other of the timescales can be ignored, in

doi:10.1088/2514-3433/abb6f0ch7

reality neither can be neglected. In addition, in practice each of the timescales is difficult to precisely determine during simulations at any given time. Thus, in what follows we will use the terms "slow spiral-in" and "self-regulated spiral-in" based on the specific definition used to mark the end of the plunge-in phase, without meaning to imply that dynamical processes or energy transport processes can be neglected in the subsequent evolution. However, please note that the terms are very much interchangeable.

Orbital eccentricity also does not provide a good criterion to define the end of the plunge. The trajectory of the companion during the plunge-in is not Keplerian, and it cannot clearly be characterized as having a particular eccentricity. In addition, it is not clear if the post-plunge binary should have a negligible eccentricity, though during a slow spiral-in it tends to decrease.

On the other hand, the rate of change of orbital energy in the system consisting of the core and the companion, in any fixed reference frame, can be measured relatively well from three-dimensional simulations. In a way similar to the case of a non-point gravitational potential in galactic dynamics, let us consider the radial period, which is the time necessary to travel from pericenter to apocenter and back. Here we will use the radial period for the orbital period $P_{\rm orb}$. This makes the following criterion useful:

$$|\dot{E}_{\rm orb}P_{\rm orb}/E_{\rm orb}| \leqslant 0.01. \qquad (7.1)$$

The criterion above is not the only one possible, but it is easy enough to obtain numerically in any 3D simulation and hence can serve as a comparison between codes.

Independently of the chosen criterion, it is useful to report the measured timescales of the envelope and of orbital shrinkage as functions of time. Indeed, if the numerical method has been created to model a slow spiral-in, but the physical situation considered is during the plunge-in, the outcome can be expected to be improper. We remind the reader that pre-common-envelope timescales do not provide good criteria for determining the beginning of the slow spiral-in. In particular, note that timescales computed using the initial donor structure do not accurately indicate the location where slow spiral-in should begin.

7.2 Which Processes Are Important?

7.2.1 Energy Transport in the Envelope

During the slow spiral-in, the envelope can transfer away energy it receives. An implicit assumption is that the mechanical energy it receives is reprocessed into heat, and that heat is being transferred throughout the envelope by radiation or convection.

Accordingly, a numerical method that models a slow spiral-in must include radiative transfer and convective energy transport (conduction as energy transport can be neglected in case of a CEE, since the envelope is usually nondegenerate).

Some three-dimensional codes, given sufficient resolution, may recover the onset of turbulent convection (Ohlmann et al. 2016). However, at least for the purpose of common envelope evolution, there has not yet been a simulation that can demonstrate true convective energy transport from the bottom of the common envelope to its surface. Three-dimensional codes with radiative transport exist, but have not yet been applied to the slow spiral-in. Overall, modeling the slow spiral-in using a three-dimensional code is computationally expensive and not yet feasible. This is true even if no non-advective energy transport mechanisms are taken into account. To reiterate, the main reason for hydrodynamical codes to be expensive is that the time step has to be as small as a fraction of the diminished binary orbital period, while the envelope's dynamical timescale can easily exceed the new orbital period by four orders of magnitude (see examples in Ivanova & Nandez 2016). This means that the number of time steps required to evolve the system through multiple expanded envelope timescales will be of order 10^5 or more (Section 4.3.1).

A typical one-dimensional stellar-evolution code generally includes approximations to both mechanisms of energy transport. Specifically, convective energy transport is included using mixing length theory (MLT; e.g., as in the approximation in Böhm-Vitense 1958), and radiative transfer is included in the diffusion approximation. However, it is possible that both approximations used in one-dimensional codes may break for the CEE situation.

Let us begin with convection. Note that both of the standard criteria for convective instability, as well as the MLT prescription, assume a one-dimensional stratified hydrostatic background. This does not hold during a CE event. Interaction between convective and pulsational motions, for example, requires a more sophisticated treatment of convection (e.g., Quataert & Shiode 2012). Further, the MLT approximation is derived under the assumption that convection is subsonic and a convective eddy can always come into pressure equilibrium with its surroundings as it moves. This requires that convection be slow and limits how much energy it can carry during a common-envelope event. For example, it has been found that to carry all energy released by hydrogen recombination, convection must become substantially supersonic, and convective eddies often must be accelerated to velocities exceeding even their local escape velocity (Ivanova et al. 2015; Ivanova 2018).

The assumption of optically-thick conditions, in local thermodynamic equilibrium, used by one-dimensional codes also limits the applicability of predictions based on them. Specifically, problems arise because we see situations when a significant part of the envelope is optically thin. First, the apparent "photospheric boundary" has a non-negligible mass, with a thickness comparable to the stellar radius and a spherical (non-planar) geometry. Second, the dominant radiative transfer may also take place in lines rather than the continuum, so the diffusion approximation breaks down in the optically thin part of the envelope. This problem was noted a long time ago for single large giants (e.g., Paczyński 1969), but there is not yet a working remedy.

7.2.2 Equation of State

As the common envelope expands and cools, the material passes through phase transitions, from an ionized to an atomic and then a molecular state. For example, hydrogen changes as $2H^+ + 2e^- \Rightarrow 2H \Rightarrow H_2$. A common envelope during a slow spiral-in may, in principle, consist of matter in all three phases at the same time. The energy released during such transitions is comparable to the kinetic energy of the atoms and needs to be taken into account. During these transitions the ratio of specific heats, $\gamma = c_P/c_V$, changes when energy is released. At the same time, the adiabatic exponent that measures the response of the pressure to adiabatic compression or expansion, $\Gamma_1 \equiv \gamma_{\rm ad} = (\partial \ln P/\partial \ln \rho)_{\rm ad}$, also changes. For example, for an ideal gas with solar composition, Γ_1 can decrease in a partial ionization zone from 5/3 to below 1.2.

Therefore, two things need to be accounted for. First, one needs to use an equation of state that includes latent heat as the material goes through the phase transition. Second, one also needs to account for changing adiabatic exponents and specific heats, and the fact that the three adiabatic exponents are usually not equal except for certain values, e.g., $\Gamma_1 = \gamma = 5/3$ (see, e.g., Stothers 2002). In principle, hydrodynamic codes can formally replace their adopted gamma-law equation of state with a tabulated equation of state. However, changes of adiabatic exponents create a more complicated problem, as was discussed in more detail in Section 4.1.3.

One-dimensional stellar codes conventionally include tables for equations of state for atomic and ionized states, and some of them include equations of state accounting for molecules. One needs to be careful, however, with how the one-dimensional code solves the hydrodynamics equations, as it may experience similar problems as three-dimensional hydrodynamic codes.

7.2.3 The Envelope's Rotation After the Plunge

The total angular momentum that the envelope has by the start of the slow spiral-in, and its distribution, may affect the drag that the envelope imposes on the companion, and hence the rate of the orbital dissipation during the slow spiral-in (for more details see Section 7.2.4). If there is non-negligible angular momentum deposition, it could also generate differential rotation in the envelope. Differential rotation results in frictional luminosity (Meyer & Meyer-Hofmeister 1979): the rotation-related kinetic energy which was injected into the envelope during the plunge-in is being converted into internal energy, but now at a much slower rate, on the envelope's viscous timescale. While this conversion of energy from one type into another does not change the overall energy budget, it affects three very important aspects of the slow-spiral-in evolution:

- where the envelope is being heated;
- on what timescale the envelope is being heated, $\tau_{\rm fric}$;
- how the rotational kinetic energy is used to eject envelope material.

These effects will be discussed below in Section 7.4.1.

7.2.4 Drag Around the Companion and the Orbit Dissipation

The drag that the envelope exerts on the companion includes both non-local and local effects. Here, we will consider three effects, although there could be more.

The first example is the non-local effect of tidal (gravitational) interactions between the companion and the entire envelope. Non-local tidal energy dissipation has been identified during three-dimensional simulations of plunge-in phases (e.g., Ricker & Taam 2008; Passy et al. 2012). However, it might be absent during the slow spiral-in (in other words, the slow spiral-in starts when non-local tidal dissipation diminishes). The rate of non-local tidal dissipation is not certain at all. Using the mass estimate for a tidal bulge in Savonije & Papaloizou (1985); the approximate orbital dissipation time of a non-eccentric binary via tidal interaction is (Ivanova 2002):

$$\tau_{\rm tid} \approx \frac{16 P_{\rm orb}}{\pi \eta_{\rm bulge}} \frac{q^4}{1+q}. \tag{7.2}$$

Here $\eta_{\rm bulge}$ is the ratio of the density of the bulge to the average density in the envelope at the same distance to the center of mass, with $\eta_{\rm bulge} \gtrsim 1$, while $P_{\rm orb}$ is the current binary orbital period. If the donor is four times more massive than the companion, $q = 4$, and we take $\eta_{\rm bulge} = 1$ (i.e., we assume that the tidal bulge is not significantly more dense that the envelope), $\tau_{\rm tid} \approx 260 P_{\rm orb}$. Recall that the plunge-in operates on a dynamical timescale, which is comparable to the initial orbital period. At the start of the plunge-in, the tidal dissipation time is much larger than the characteristic timescale for orbital shrinkage, and hence does not affect the plunge-in. However, as the binary shrinks, the tidal dissipation time can become comparable to the new (increased) dynamical timescale of the expanded envelope. We urge strong caution, however, as the tidal bulge mass estimate is made when the binary is *outside* of the envelope, not inside. The tidal effects can be expected to be negligible when the binary orbit is smaller than 1/6 of the radius within which most of the envelope mass is located (Nandez & Ivanova 2016).

The second example is the local effect of viscous drag on a sphere (e.g., Ivanova 2002), which is mainly important for nondegenerate companions (because they are much larger than compact object for a given mass). A hard sphere moving through a viscous medium, characterized by a Reynolds number $Re \gg 1$, is affected by a viscous drag force that is (see the problem at the end of §45 in Landau & Lifshitz 1959)

$$F_{\rm visc} = 12 \pi R_{\rm comp} \eta_{\rm visc} v_{\rm rel}, \tag{7.3}$$

where $v_{\rm rel}$ is the relative velocity of the companion with respect to the envelope, $R_{\rm comp}$ is the companion's radius, and $\eta_{\rm visc}$ is the shear viscosity coefficient. In the convective zone the viscosity is dominated by convective transport and hence is provided by convective eddies, so it can be estimated as $\eta_{\rm visc} = \eta_c = \alpha_c \rho v_c l$, where ρ is local density, v_c is the convective velocity, and l is the mixing length. α_c is a dimensionless parameter of order unity; for example, Meyer & Meyer-Hofmeister (1979) adopted a value of 0.5 for α_c, which assumes that convective eddies of size half the mixing length are dominant for determining the viscosity, while Bisnovatyi-Kogan (2001) quotes $\alpha_c = 1$. The characteristic dissipation timescale based just on this sphere passing through a viscous turbulent medium can be estimated as (Ivanova 2002)

$$\tau_{\rm visc} \approx \frac{1}{12\pi R_{\rm comp}\eta_{\rm visc}} \frac{q}{1+q}. \tag{7.4}$$

For a binary with a mass ratio of four, containing a secondary star with a radius similar to the Sun, in the limiting case of a non-corotating envelope when $v_{\rm rel}$ equals the orbital velocity, and with an assumed typical average viscosity of 10^{11} g cm^{-1} s^{-1} (see, e.g., Meyer & Meyer-Hofmeister 1979), the viscous spiral-in time is about 50 years.

The third example is a local effect due to gravitational polarization of the envelope material, i.e., the forward-backward asymmetry with respect to the motion of the companion inside the envelope, induced in the envelope near the companion. (Dynamical friction from a fluid background is a common problem in astrophysics; see, e.g., Ostriker 1999.) Because only the mass and not the size of the companion matters, this type of drag can be important for compact companions in CEE. This is in strong contrast with turbulent viscous drag.

For a gaseous envelope with uniform background density ρ_0, the magnitude of the dynamical drag force on motion at velocity v_∞ can be estimated as (Ostriker 1999):

$$F_{\rm DF} \approx C_{\rm d} \times 4\pi \frac{\left(GM_{\rm comp}\right)^2 \rho_0}{v_\infty^2}, \tag{7.5}$$

where $C_{\rm d}$ is the drag coefficient. From perturbation theory $C_{\rm d}$ is expected to be equal to $\mathcal{M}^3/3$ for relative motion with $\mathcal{M} \ll 1$ and become greater than one for $\mathcal{M} \gg 1$.

Using a numerical wind-tunnel experiment, MacLeod et al. (2017) studied the dependence of $C_{\rm d}$ for cases involving non-uniform background density. Unlike accretion efficiency, which in the same study was found to be significantly lower than provided by analytic predictions like ideal Hoyle–Lyttleton accretion, the drag coefficient they numerically obtained is about one. It was found to rise in flows with steeper density gradients, and to be in the range 0.3–6.5 (see their Figure 10, and for the ratio of accretion coefficient $C_{\rm accr}$ to drag coefficient $C_{\rm d}$, see Figure 3.3).

The dissipation timescale due to drag forces can be estimated as the ratio of the orbital energy to the power with which drag removes energy from the orbit. Let us measure the companion motion relative to the local envelope material in units of its Keplerian orbital velocity $v_\infty = f_{\rm k} v_{\rm orb}$, and let $\rho_{\rm orb}$ be the density near the orbit. Using Equation (7.5), this gives:

$$\tau_{\rm DF} = \left| \frac{E_{\rm orb}}{F_{\rm DF} v_\infty} \right| = \frac{1}{4\pi C_d G} \times \frac{f_k}{\rho_{\rm orb} P_{\rm orb}} \times \frac{q^2}{1+q}. \tag{7.6}$$

For a given binary mass ratio, the largest uncertainty in this expression is the density in the vicinity of the orbit. For example, assume that the slow spiral-in orbit is at $a \sim 10^{11}$ cm, and the binary consists of a $2M_\odot$ donor with a $0.4M_\odot$ core and a $0.5M_\odot$ companion. If the local density is no larger than 10^{-6}g cm^{-3}, $\tau_{\rm DF}$ can be as short as a dozen years. However, by the start of the slow spiral-in envelope expansion may have reduced the density by several orders of magnitude, making the characteristic timescale correspondingly larger.

Indeed, analysis of available three-dimensional simulations after the end of the plunge-in has shown strong decompression around the transitional binary orbit, up to almost complete envelope evacuation from this region (Ivanova & Nandez 2016). This makes tidal and viscous drags numerically negligible. However, if this material is not unbound it also presumably leads to future envelope re-collapse, which would change the steady course of common envelope evolution that is usually assumed for a slow spiral-in. Overall, there is no reason to expect this drag to be constant during the entire slow-spiral-in; it depends on the instantaneous state of the envelope.

The total drag on the companion from all sources causes the orbit to shrink and heats the envelope, so the rates of these processes are linked. While we have a useful framework for estimating the rates of these processes as functions of local conditions, the precise manner in which the envelope is heated is still not well-established. Moreover, since conditions near the companion depend on the state of the envelope after the plunge-in, the drag rate and orbital dissipation timescale $\tau_{\rm drag}$ are also uncertain. Of these effects, the largest uncertainty is associated with the non-local tidal drag.

7.2.5 Accretion

As discussed in detail in Section 3.3.1, accretion can provide significant input only in cases when the slow spiral-in phase lasts a long time. We discussed there how the accretion rate can be much smaller than expected from BHL accretion theory and the Eddington-limited rate. We also mentioned in Section 7.2.4 that the envelope can be evacuated from the companion's orbit, and hence accretion may not even take place.

Let us examine the following limiting case: if the accretion is as high as the Eddington rate, then it becomes energetically important if the slow spiral-in lasts longer than

$$\tau_{\rm accr} = \frac{E_{\rm bind}}{L_{\rm Edd}} = \frac{GM_{\rm env,si}M_{\rm d,si}}{\lambda_{\rm si}R_{\rm si}}\frac{\kappa_{\rm si}}{4\pi GcM_{\rm comp}} \simeq 2400\ {\rm yr}\ \frac{q_{\rm si}}{\lambda_{\rm si}}\frac{\kappa_{\rm si}}{{\rm cm^2 g^{-1}}}\frac{M_{\rm env,si}}{M_\odot}\frac{R_\odot}{R_{\rm si}}, \quad (7.7)$$

where $\kappa_{\rm si}$ is the value of the opacity near the companion, $M_{\rm d,si}$ and $M_{\rm env,si}$ are the masses of the donor and of the common envelope during the slow spiral-in, $R_{\rm si}$ is the radius of the common envelope during the slow spiral-in, and $q_{\rm si} = M_{\rm d,si}/M_{\rm comp}$. $\lambda_{\rm si}$ is a parameterization for the common envelope's energy during the slow spiral-in; it is not the same as the donor's initial λ. From Equation (7.7), if the envelope has expanded to about $100R_\odot$, the opacity near the companion is provided by Thomson scattering, and the companion accretes at the Eddington limit, accretion can be a major energy input for a slow spiral-in that lasts for about 30 years or longer.

7.2.6 Roche-lobe Overflow by the Companion

Under some circumstances the slow spiral-in may continue until the companion overfills its Roche lobe. In that case the direction of mass transfer would be reversed, i.e., from the companion to the core of the donor. The timescale for mass transfer would be mainly a function of the drag forces due to the common envelope, which would continue to operate. This timescale would be ~10–1000 yr, which is faster than the thermal time but slower than the dynamical timescale of the companion. This could continue until the companion was completely destroyed (Ivanova 2002).

More speculatively, depending on the evolutionary states of the donor and the companion, the thick hydrogen-rich stream from the companion may penetrate into the hot core as deeply as the helium-burning zone (Ivanova et al. 2002). If the stream penetrated as deeply as the helium-burning shell, it could lead to very rapid hydrogen burning with a high energy generation rate, leading to an immediate end of the slow spiral-in via so-called explosive common-envelope ejection. This might lead to unusual nucleosynthesis, e.g., enhancement with *s*-process elements, and might potentially be one of the paths to create long-gamma-ray burst progenitors (Podsiadlowski et al. 2010).

7.2.7 Self-consistency and Relevant Timescales

Note that an "idealized" slow spiral-in is a phase during which the expanding envelope radiates away most of the energy it receives from the orbiting core and companion. This requires a balance between the surface luminosity and the rate at which mechanical energy is injected into the envelope. In the absence of a self-regulatory mechanism, the necessary fine-tuning would make this behavior unlikely. The surface luminosity can only change on the thermal adjustment timescale of the envelope, $\tau_{\rm adj,CE}$, while the rate of mechanical energy injection changes on the drag timescale, $\tau_{\rm drag}$. If $\tau_{\rm drag} < \tau_{\rm adj,CE}$, then changes in the structure of the envelope always lag behind the changing instantaneous energy input from the drag luminosity. In that case, self-regulation is no longer possible.

7.3 Transition from the Plunge to the Slow Spiral-in

Some three-dimensional simulations have been continued into the slow spiral-in stage. Using simulations with low-mass giant donors, Ivanova & Nandez (2016) identified the following results as potentially useful for initializing one-dimensional slow spiral-in simulations. Note that they may not apply to massive donors.

- By the start of the slow spiral-in, the envelope has lost a substantial fraction of its mass during the mechanical "plunge-in" ejection phase. This result is generally supported by all hydrodynamic simulations of the plunge-in, even if the simulation does not reach the slow spiral-in phase. While initial conditions chosen for simulations are not necessarily an unbiased representation of all CE phases, simulations of slow spiral-in phases should account for mass-loss during the plunge.
- The remaining matter in the bound envelope has a specific entropy profile that is almost unchanged as compared to the pre-CE Lagrangian entropy profile.
- The gravitational potential in the envelope surrounding the binary and the common envelope deviates from the point-mass approximation.

- Most of the pre-plunge total angular momentum is lost from the system with the material ejected prior the start of the slow spiral-in. The fraction of the initial angular momentum transferred to the bound envelope is likely to be low.
- The 3D simulations in Ivanova & Nandez (2016) that included a tabulated equation of state resulted in the establishment of recombination outflows during the slow spiral-in. These outflows started to remove the envelope on a timescale between the thermal and dynamical timescales of the expanded envelope. Hence 3D simulations that are carried past the beginning of these outflows (when they occur) must include an appropriate equation of state.

7.4 What Have We Learned from One-dimensional Simulations?

One-dimensional slow spiral-in simulations can be divided into two kinds by the way in which the evolution of the donor's envelope in treated. In the first type one adopts some constant rate of heating, which effectively imposes a rate of decay for the binary orbit. The second type of simulation is often called a *"self-consistent"* simulation. As we will discuss below, each of the types of one-dimensional simulation provides some useful physical *insights*, but neither so far is close to *reality*.

In each case, orbital energy is deposited into the common envelope by adding a source of heat. Comparison with three-dimensional simulations has shown that this leads to substantially higher local entropy generation (an "entropy bubble") than is observed in three-dimensional simulations (see more details in Section 6.4). At the same time, matter is gravitationally accelerated by the companion much less than in the three-dimensional case. One-dimensional simulations may incorrectly predict which part of the envelope can be ejected, since the recombination radius (see Section 8.3) will be different when the recombining matter has a higher entropy.

7.4.1 Constant Heating

The following assumptions are standard for this type of simulation:

- An overall timescale ($\sim(1-100)\tau_{\rm th,CE}$) is imposed for the slow spiral-in. Note that the rate of inspiral is typically not assumed to be constant, but the rate of heating is.
- Energy is deposited into the envelope as thermal energy at a rate equal to the rate of change in gravitational potential and orbital energies.
- The heating is distributed at an assumed set of locations. These can, e.g., be fixed in mass or in radius near the assumed orbit, or at the base of the envelope.

The total amount of energy injected is based upon the amount of orbital energy likely to be available:

$$\Delta E \sim 2\times 10^{48}\ {\rm erg}\ \frac{M_{\rm d}}{M_\odot}\frac{M_{\rm comp}}{M_\odot}\frac{R_\odot}{a}. \tag{7.8}$$

Here a is a notionally plausible characteristic orbital separation during some part of the slow spiral-in.[1] The range of timescales assumed for the duration of this phase varies from approximately one year to ~1000 yr. A typical orbital separation during a slow spiral-in can be between one and hundreds of solar radii. The rate of the energy injection (heating luminosity) for a binary with a $2M_\odot$ donor and a $0.5M_\odot$ companion can vary widely, from $L_{\rm heat} = 6 \times 10^{35}$ to 6×10^{40} erg s^{-1}. For this process to be sustained, the heating luminosity cannot exceed the plausible maximum surface luminosity, leading to a minimum heating timescale during a slow spiral-in of roughly one year.

The donor is expected to reach equilibrium with the heating luminosity during the slow spiral-in stage. Note that the maximum plausible luminosity is likely dictated by the Eddington luminosity, which is

$$L_{\rm Edd} = \frac{4\pi c G M_*}{\kappa} \approx 5 \times 10^{37}\ {\rm erg\ s^{-1}}\ \frac{1}{\kappa}\frac{M_*}{M_\odot} \tag{7.9}$$

for a star with mass M_*. Typically, the Eddington luminosity expression assumes the opacity to be the Thomson scattering opacity of a typical stellar mixture, $\kappa = \kappa_{\rm TS} = 0.19(1 + X)$. Then $L_{\rm Edd} \approx 1.5 \times 10^{38} M_*/M_\odot$. However, an expanded cold common envelope *can have a surface opacity* κ *which is much lower than the Thomson-scattering opacity*, hence the surface luminosity during a spiral-in can be higher than its Thomson scattering-limited Eddington luminosity (see e.g., Ivanova et al. 2015).

Importantly, the outcome of constant-heating studies is not simply a function of how much total energy is deposited, as is usually assumed, but also *where* and *how rapidly* it is deposited. In other words, it depends on the local heating rate per unit mass. The deposition of energy in a more narrow strip around the binary orbit, as well as deposition of the same amount of energy but more rapidly, leads to more energetic ejections or makes the envelope more unstable than slower deposition or deposition spread over a larger part of the envelope (Ivanova et al. 2015).

The donor expands until it can radiate away all the heating luminosity. However, a high specific rate of local heating can cause the rate of expansion of the envelope to exceed its escape velocity. For a given total heating rate, a higher specific local heating rate can cause the envelope to escape completely, while a lower specific local heating rate can leave the donor in a thermally expanded state while the spiral-in continues.

As the envelope expands, a substantial part of the envelope's hydrogen and helium recombines. One consequence is that in a partial ionization zone the local first adiabatic index, defined as

$$\Gamma_1 = \left(\frac{\partial \ln P}{\partial \ln \rho}\right)_{\rm ad} \tag{7.10}$$

[1] Welcome to astrophysics. The choices made in this area historically have not been well-motivated.

can become as small as 1.1. This affects the dynamical stability of the envelope, which depends on the pressure-weighted, volume-averaged value of Γ_1 (Ledoux 1945; Stothers 1999):

$$\langle\Gamma_1(m)\rangle = \frac{\int_m^{M_d} \Gamma_1 P\, dV}{\int_m^{M_d} P\, dV}. \tag{7.11}$$

If $\langle\Gamma_1(0)\rangle < 4/3$, the whole star is dynamically unstable. It has been argued for cool giants that if $\langle\Gamma_1(m_{\rm env})\rangle < 4/3$ the envelope material outside the enclosed mass $m_{\rm env}$ is unstable (Lobel 2001).

Simulations with constant heating assuming hydrostatic equilibrium have found that a substantial part of the heated envelope, in its stationary state, has $\langle\Gamma_1\rangle < 4/3$ (Ivanova et al. 2015). Simulations with constant heating and including the dynamical terms in the equation of motion have demonstrated that the heated envelope initiates large-amplitude pulsations (Clayton et al. 2017). During the pulsations the star evolves to the right of the Hayashi line and could be observed as a very cold variable star with a pulsation period of several years (for more details, see Section 10.6).

7.4.2 "Self-consistent" with Feedback

This type of simulation is not truly self-consistent. It adopts some model of how the orbital decay depends on the immediate conditions inside the common envelope, as well as some model for the orbit-dependent location where heat from the orbital decay is injected into the envelope.

These studies were pioneered by Taam et al. (1978) and Meyer & Meyer-Hofmeister (1979). During the 40 years since then, numerous independent studies have been performed, usually intending to address specific cases. Below, we attempt to list which specific physics was added for the first time (i.e., we do not list a study if it used qualitatively similarly approximate physics later). Unlike the "constant heating" approach, most "self-consistent" studies have not provided systematic insights into common envelope physics, but have rather demonstrated interesting plausible specific outcomes.

Physical processes that have been included to produce the feedback include:

- Angular momentum transfer from the orbit to the envelope, leading to differential rotation of the envelope (Taam et al. 1978; Meyer & Meyer-Hofmeister 1979).
- Heating due to the frictional luminosity inside a differentially rotating envelope (Taam et al. 1978; Meyer & Meyer-Hofmeister 1979).
- Tidal drag from the envelope on the companion (Taam et al. 1978; Meyer & Meyer-Hofmeister 1979).
- Gravitational capture (accretion-related) drag on a degenerate companion (Taam et al. 1978).

- Viscous drag from the envelope on a nondegenerate companion (Ivanova 2002).
- Change in the donor's gravitational potential due to the secondary (Ivanova 2002).
- Response of a main sequence donor to accretion while inside the common envelope (Hjellming & Taam 1991).
- Chemical element mixing due to spin-up of the envelope (Ivanova 2002).
- Nuclear physics during companion–core interaction (Ivanova et al. 2002).

Note that in all these cases the added physics is very simplified and does not necessarily render reality; effects of such simplifications were discussed elsewhere earlier in this Chapter.

One-dimensional codes have often encountered problems when the envelope's matter becomes unbound, especially when a significant density inversion is present. However, with time the numerical stability of one-dimensional codes has improved substantially, as the numerical methods used to obtain convergence between iterations have been changed. One of the most recent one-dimensional "self-consistent" common-envelope simulations demonstrated the removal of most of the hydrogen-rich envelope of a massive donor star (Fragos et al. 2019).

References

Bisnovatyi-Kogan, G. S. 2001, Stellar Physics. Vol. 1: Fundamental Concepts and Stellar Equilibrium (Berlin: Springer)
Böhm-Vitense, E 1958, ZA, 46, 108
Clayton, M., Podsiadlowski, P., Ivanova, N., & Justham, S. 2017, MNRAS, 470, 1788
Fragos, T., Andrews, J. J., Ramirez-Ruiz, E., et al. 2019, ApJL, 883, L45
Hjellming, M. S., & Taam, R. E. 1991, ApJ, 370, 709
Ivanova, N. 2002, DPhil thesis, Balliol College, Oxford
Ivanova, N. 2018, ApJL, 858, L24
Ivanova, N., Justham, S., & Podsiadlowski, P. 2015, MNRAS, 447, 2181
Ivanova, N., Justham, S., & Podsiadlowski, P. 2015, MNRAS, 447, 2181
Ivanova, N., & Nandez, J. L. A. 2016, MNRAS, 462, 362
Ivanova, N., Podsiadlowski, P., & Spruit, H. 2002, MNRAS, 334, 819
Landau, L. D., & Lifshitz, E. M. 1959, Fluid Mechanics (London: Pergamon)
Ledoux, P. 1945, ApJ, 102, 143
Lobel, A. 2001, ApJ, 558, 780
MacLeod, M., Antoni, A., Murguia-Berthier, A., Macias, P., & Ramirez-Ruiz, E. 2017, ApJ, 838, 56
Meyer, F., & Meyer-Hofmeister, E. 1979, A&A, 78, 167
Nandez, J. L. A., & Ivanova, N. 2016, MNRAS, 460, 3992
Ohlmann, S. T., Röpke, F. K., Pakmor, R., & Springel, V. 2016, ApJL, 816, L9
Ostriker, E. C. 1999, ApJ, 513, 252
Paczyński, B. 1969, AcA, 19, 1
Passy, J.-C., De Marco, O., Fryer, C. L., et al. 2012, ApJ, 744, 52
Podsiadlowski, P., Ivanova, N., Justham, S., & Rappaport, S. 2010, MNRAS, 406, 840

Quataert, E., & Shiode, J. 2012, MNRAS, 423, L92
Ricker, P. M., & Taam, R. E. 2008, ApJL, 672, L41
Savonije, G. J., & Papaloizou, J. C. B. 1985, in NATO Advanced Science Institutes (ASI) Series C, Vol. 150, Interacting Binaries, ed. P. P. Eggleton, & J. E. Pringle (Berlin: Springer), 83
Stothers, R. B. 1999, MNRAS, 305, 365
Stothers, R. B. 2002, ApJ, 581, 1407
Taam, R. E., Bodenheimer, P., & Ostriker, J. P. 1978, ApJ, 222, 269

Natalia Ivanova, Stephen Justham and Paul Ricker

Chapter 8

Mechanisms of Mass Ejection

The removal of a common envelope does not necessarily proceed by one single episode of mass outflow. As it is currently understood, several different phases of mass ejection could take place during the same common-envelope event. We discuss below the four phases identified so far, driven by different physics. Figure 8.1 shows a calculation that demonstrates all four types, although in practice all may not be present during a given common-envelope episode. The first phase of ejection is the initial ejection and takes away a significant fraction of the angular momentum. The second phase takes place during the plunge and is driven mainly by orbital energy deposition. The third phase is driven by recombination energy and takes the form of nearly-steady outflows. The final ejection is linked to the overall long-term instability of the envelope. This instability erupts through the interplay between the fallback of envelope material and its reionization. It can take the form of either a partial mass ejection, in which case it is referred to as a shell-triggered instability, or complete ejection, in which case it is a delayed dynamical ejection.

8.1 Initial Ejection

The initial ejection is the removal of the very outer layers of the original envelope and takes place before the plunge-in starts. It is seen in all simulated common-envelope evolutions (CEEs), including those that end with the merger of the two stars. The ejecta carry away a significant fraction of the initial angular momentum, but little mass is removed.

The amount of mass that is ejected prior the plunge-in, $\delta M_{\rm ej,pre}$, can be estimated by analogy with the energy formalism. One can compare the orbital energy release $\delta E_{\rm orb}(r)$ for inspiral to a radial coordinate r to the local binding energy $\delta E_{\rm bind,end1D}(r)$ of the envelope outside that radius (Ivanova et al. 2013). We define the local energy imbalance, $\Delta E(r)$, as

$$\Delta E(r) \equiv \delta E_{\rm bind,end1D}(r) + \delta E_{\rm orb}(r). \quad (8.1)$$

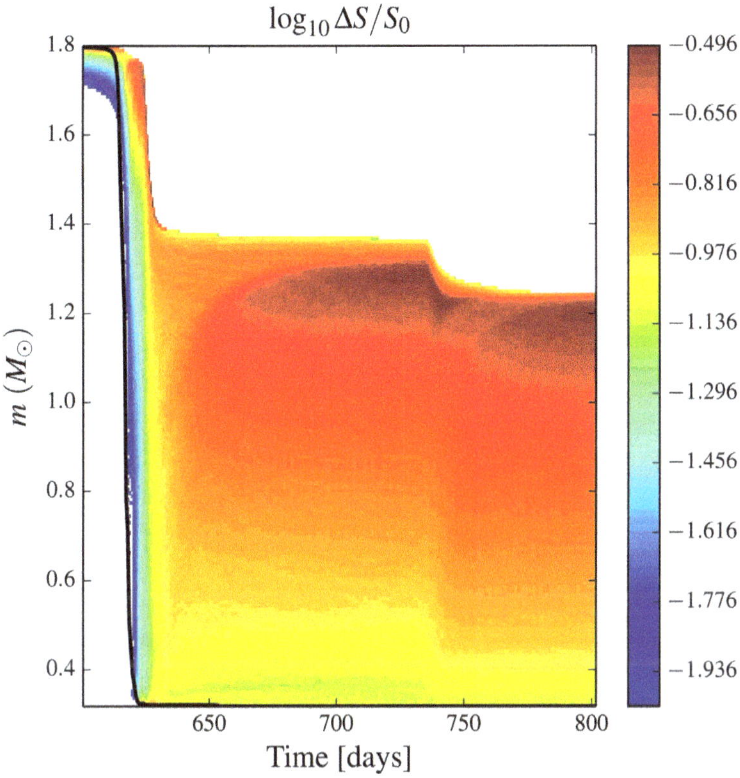

Figure 8.1. The results of a three-dimensional simulation that did not completely eject the envelope, but which demonstrates four types of ejection. The initial phase of ejection lasts up to about 620 days in this plot; between about 630 and 640 days the plunge-in ejecta are released; after about 640 days the recombination outflows occur; and between about 735 and 745 days the "shell-triggered" ejecta are released. This is a common-envelope event involving a low-mass red giant with a mass of 1.8 $M_\odot$, a core mass of 0.318 $M_\odot$, and an initial radius of 16.3 $R_\odot$, with a point-like companion of 0.15 $M_\odot$. The black solid curve shows the location of the companion with respect to the mass coordinate of the donor star. The color demonstrates the logarithmic change of the entropy, shown as the ratio of the cumulative increase of the specific entropy from the start of the simulation with respect to its initial value. The color scale approximately covers a range from a 1% increase to a 30% increase in specific entropy. A white color below or around the orbit means no change in entropy. White space in the top part means that this part of the mass of the envelope is currently unbound and is removed from the plot. The entropy increase mainly comes from recombination. This figure is reproduced from Ivanova & Nandez (2016). © 2016 The Authors. CC BY.

The standard energy formalism considers the moment when $\Delta E(r)$ becomes positive near the core. However, when the orbit starts to shrink, $\Delta E(r)$ is initially positive as well; see Figure 8.2. The ejected mass $\delta M_{\rm ej,pre} = M_{\rm don} - m(r_{\rm pre})$ can be found by finding the radial coordinate value $r_{\rm pre}$ at which $\Delta E(r_{\rm pre}) = 0$. Ivanova & Nandez (2016) compared the pre-plunge ejected mass $\delta M_{\rm ej}$ measured in three-dimensional simulations with the value $\delta M_{\rm ej,pre}$ obtained from a one-dimensional stellar model that was used to initialize the simulations. Values found from the two methods agree

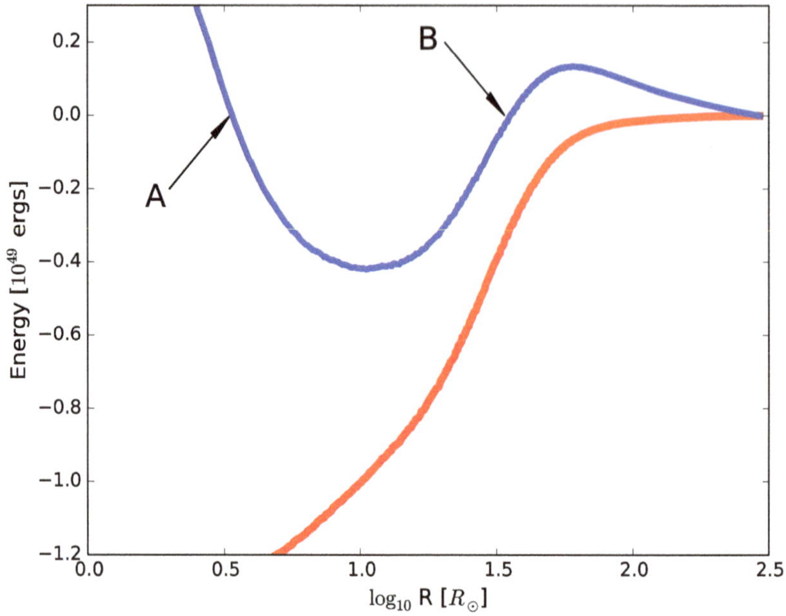

Figure 8.2. Energy balance for simplified inspiral of a 10 $M_\odot$ companion into the envelope of a 20 $M_\odot$ giant with a radius of 300 $R_\odot$. The blue curve shows the local value ΔE of the local energy imbalance under the assumption that the companion has a mass of 10 $M_\odot$. The standard energy formalism predicts that this system has to shrink so that at least the envelope above 3.4 $R_\odot$ (mass coordinate 6.9 $M_\odot$) is ejected (this location is indicated with the arrow A). However, $\Delta E(r)$ is positive everywhere for radii above ~ 35 $R_\odot$, or above the mass coordinate $m \sim 16\ M_\odot$; this location is indicated with the arrow B. For comparison, the red curve shows the local value of the envelope energy including internal energy terms, integrated from the surface.

within a factor of two. (We stress the caveat that this comparison was performed for low-mass giant donors.)

In these simulations and others the envelope was not spun up significantly during the initial mass ejection. If there is no spin-up in reality, then this initial ejection should remove an amount of angular momentum that corresponds to relocation from the initial orbit to $r_{\rm pre}$, which itself does not exceed the initial Roche-lobe radius of the donor, $r_{\rm don,RL}$. The orbital angular momentum changes when inspiraling from a separation a to a separation $r_{\rm don,RL}$ by a factor of $(r_{\rm don,RL}/a)^{3/2}$. For example, for a binary system with an initial mass ratio between the donor and the companion of 4 to 1, the angular momentum taken away by this ejection can be approximately 60% of the initial orbital angular momentum. To some extent, it can be said that the special feature of the driving mechanism behind this ejecta is the orbital angular momentum deposition, even though the matter is evacuated from the binary system using the initial energy excess.

8.2 Dynamical Plunge-in Ejection

This is the most well-known type of ejection and is usually expected to take place during a common-envelope event that results in the formation of a close binary. The main source of energy for envelope removal is orbital energy. These ejecta experience the largest mass-loss rate and likely also the fastest outflows; see Figures 8.1 and 8.3.

This type of ejection takes place in all three-dimensional simulations that model successful close binary formation. Up to now, no three-dimensional simulation has demonstrated that this ejection can remove the entire envelope, and only between 25% and 50% is typically ejected (Ricker & Taam 2012; Ivanova & Nandez 2016). As has been discussed elsewhere in this book (e.g., Section 6.3), this occurs because once the envelope expands, the drag on the companion decreases, and the orbital shrinkage slows down. Consequently, as orbital energy is not deposited any longer, the continuation of the purely dynamical ejection becomes impossible.

The counterpart of this type of ejection in one-dimensional simulations is a dynamical departure of the envelope that is observed when the rate of energy

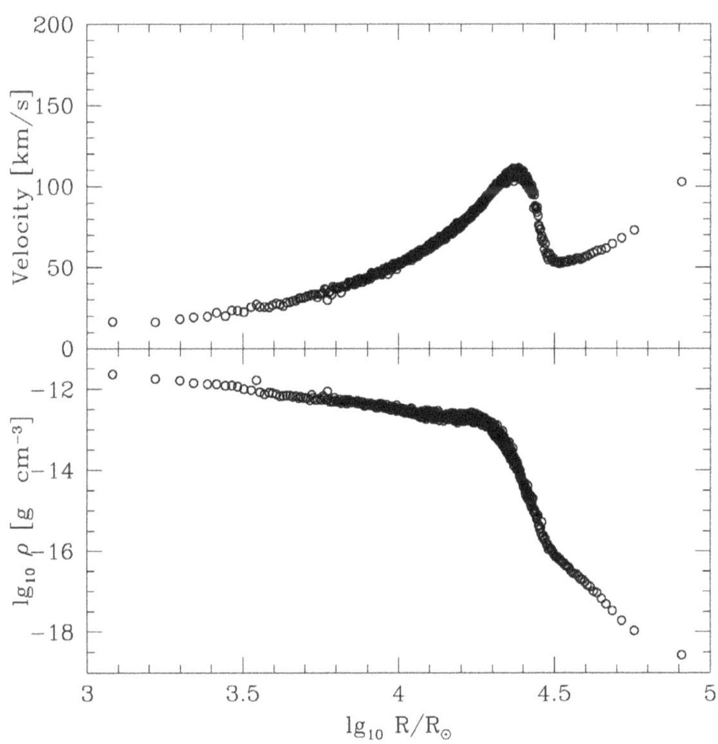

Figure 8.3. Velocity and density profiles for one typical CE ejection, shown as mass-weighted averages over all directions, for a CE simulation of an initial binary consisting of a 1.2 $M_\odot$ red giant with a 0.32 $M_\odot$ core and a 0.32 $M_\odot$ white dwarf companion. The top panel shows velocity, while the bottom panel shows density. The profile is shown 800 days after the start of the plunge. No SPH fluid particles are left within the inner 1000 $R_\odot$. The low-density initial ejecta here can be identified as the material located in the figure at log $R/R_\odot > 4.5$. The dynamical ejecta correspond to the velocity peak at log $R/R_\odot \approx 4.4$. To the left of the peak, at low velocity, is material from the recombination outflows. For comparison, the initial escape velocity from the surface of the donor is ≈ 120 km s^{-1}. This figure is reproduced from Ivanova & Nandez (2018).

deposition exceeds the donor's Thomson scattering-limited Eddington luminosity by a factor of a few (Ivanova et al. 2015; Clayton et al. 2017). We remind the reader that the Eddington luminosity is a function of opacity, and in cold envelopes, the opacity can be lower than the Thomson value (see also Chapter 7 and Equation (7.9)). In principle, as was discussed in Chapter 7, in evaluating such models one should check if the one-dimensional energy deposition is introduced directly into bulk kinetic energy or via thermal heating, as this difference is expected to affect how much material can be ejected. However, the characteristics of the ejecta look rather similar between the one- and three-dimensional cases, with velocities at the time of ejection exceeding the initial surface escape velocity by a factor of two or more.

8.3 Recombination Outflows

If after the plunge-in, expansion, and ejection, the remaining bound envelope is expanded enough for hydrogen to begin recombining, a fraction of the envelope can be removed by relatively steady "recombination outflows." For recombination outflows to be useful in removing the envelope, the radius to which the envelope has to expand during the plunge-in is found by comparing the local binding energy of the envelope and the remaining stored recombination energy (Ivanova & Nandez 2016). The combined stored helium and hydrogen recombination energy can be found using Equation (3.8). If we treat the envelope as a virialized ideal gas with $\gamma_{ad} = 5/3$, the binding energy is half the potential energy, or $-Gm_{grav}/2r$, where m_{grav} is the mass to which the envelope is still bound. In this case, for recombination outflows to be successful, the dynamical expansion should have lifted the not-yet-recombined envelope to a radius $r_{rec}/R_{\odot} \gtrsim 65 m_{grav}/M_{\odot}$. This estimate does not take into account the envelope material's kinetic energy. There are also other caveats. For instance, some recombination could have taken place before the envelope was dynamically expanded to this radius. On the other hand, the envelope could initially have had an entropy that was too high, preventing it from recombining when expanded to this radius. And after all, the dynamical plunge-in has to be powerful enough to be able to lift the envelope to this radius.

Once the envelope is sufficiently expanded, the recombination front can stall, remaining at a fixed radius while envelope gas flows through it and is accelerated by the release of recombination energy. For example, a 1D model of a low-mass CEE (Figure 8.4) shows the peak singly-ionized helium abundance remaining at a roughly constant radius throughout the slow spiral-in. This radius depends on the entropy of the envelope (see Figure 11 in Ivanova & Nandez 2016). The velocity of the ejecta is relatively small (as compared with the plunge-in ejecta), with the kinetic energy equal to a fraction of the recombination energy. Since the total specific recombination energy for all stages of hydrogen and helium is about 1.3×10^{12} erg g^{-1}, the outflow velocity should be below 36 km s^{-1}. The velocity of the recombination outflows in Figure 8.3 demonstrates that this estimate is not perfect, but in this case it appears consistent within a factor of three. This has been found to be the final ejection episode in all simulations for which a binary is formed at the end of the

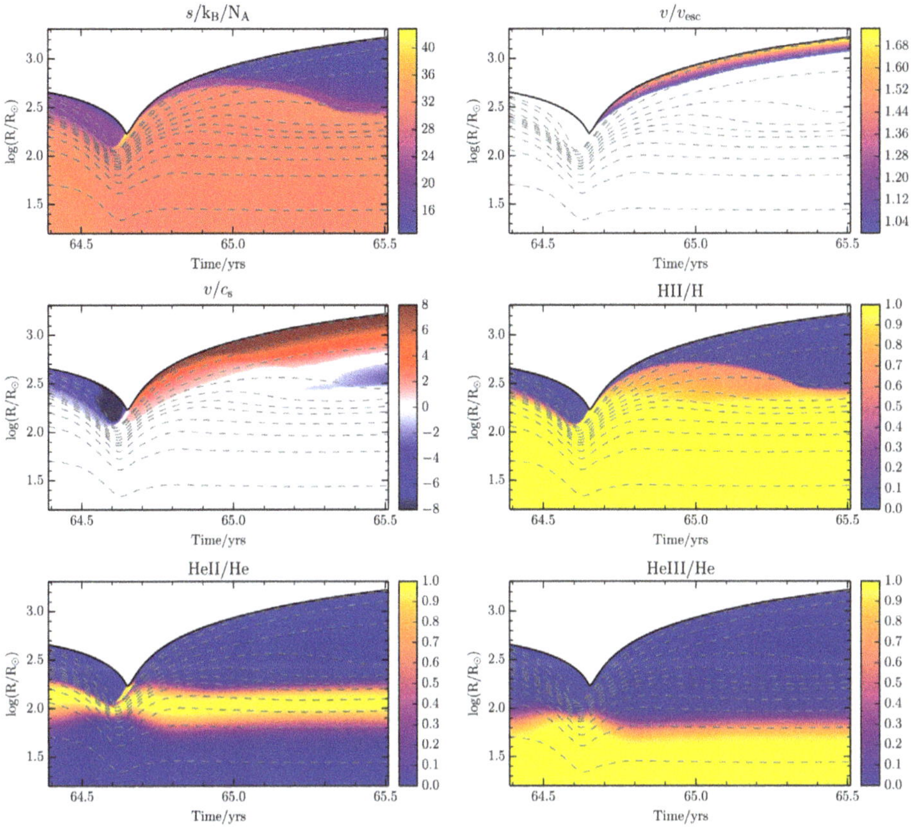

Figure 8.4. The first mass ejection displayed by a model of a 1.6 $M_{\odot}$ giant star, with an initial radius of 100 $R_{\odot}$, which is heated at a uniform specific rate throughout the convective envelope at a total heating rate of 1.7×10^{45} erg yr^{-1}. This heating regime leads to pulsations in 1D. The plot shows one of the pulsations, with a focus on the time interval near the contraction phase. The panels show dimensionless specific entropy per mole, the ratio of velocity to local escape velocity for regions where this ratio is above 1, the ratio of velocity to local sound speed for supersonic regions, and the relative proportions of ionized hydrogen and singly and doubly ionized helium. Also shown are contours containing 100% in black, and 99, 98, 95, 90, 85, 80, 75, 70, 60, 50, 40, and 30 percent in dashed gray, of the total mass of the model. This figure is reproduced from Clayton et al. (2017). © 2017 The Authors. CC BY.

common envelope event, although this picture has been established so far only for low-mass donors.

8.4 Shell-triggered Ejections and Delayed Dynamical Ejection

Recombination-mediated ejection can occur in a dynamical fashion. This is different from the steady outflows described above. The envelope material that was expanded but not ejected during the dynamical plunge-in remains bound in the sense that the sum of its potential, kinetic, and thermal energy is negative. Individual mass elements in this material follow roughly ballistic trajectories that converge as they fall back onto more slowly moving interior material. The fallback of this bound material onto the material that was not pushed out creates a shock that can suddenly, but with a time delay with respect to the end of the plunge-in, compress and reionize the infalling material, after which it re-expands, recombines, and is partially ejected (see Figure 8.4).

We can provide a tentative interpretation of the available computational experiments. Without reionization, a blob of gas dropped (figuratively) onto concrete would compress and heat on impact, after which it would re-expand and flow back up without reaching its initial height; some of the initial potential energy would be transformed into internal energy. With reionization, the blob would be kept compressed longer. In the context of the infalling bound envelope, this allows the gas layers below the shock to behave more like a trampoline than like concrete. The gas below the shock is compressed by the infalling material, but it requires time to re-expand, and once it does so it pushes the reionized, infallen gas to re-expand and recombine, at which time some of that material gains enough bulk kinetic energy to be ejected. Hence recombination energy here is being used less as a source of energy and more as a confinement mechanism that allows the infalling material to take energy from the gas interior to it. Note that the total energy budget (if not accounting for radiative energy losses, which are larger during steady recombination outflows, due to their longer timescale) is the same as with recombination outflows, but the mass loss rate can speed up during short periods of time.

Three-dimensional simulations by Ivanova & Nandez (2016) find phenomenology somewhat similar to that observed in one-dimensional calculations (Clayton et al. 2017). In those three-dimensional simulations, the overlapping of the layers traveling on their ballistic trajectories leads to a sudden post-plunge mass outflow, quickly removing $\sim 0.1\ M_{\odot}$. The three-dimensional simulations did not proceed long enough to encounter any subsequent mass ejections, but the one-dimensional simulations by Clayton et al. (2017) found quasi-periodic episodes of mass ejection, each ejecting up to $0.1\ M_{\odot}$, which might remove the entire envelope over the duration of the slow spiral-in phase.

Why are the ejections (quasi-)periodic? As the envelope has developed hydrogen and helium recombination profiles that together may span the entire envelope, the envelope's pressure-weighted, volume-averaged Γ_1 can become smaller than 4/3, causing the envelope to become dynamically unstable (see Section 7.4.1). In the case of constant heating, the envelope expands until its surface luminosity becomes comparable to the heating luminosity, and then it begins to pulsate around its "equilibrium" luminosity. Energy stored during the compression periods helps to remove the outer layers during the expansion phases. The ejections are (quasi-) periodic, as the periods of the pulsations are not exactly the same from peak to peak. Even at heating rates that do result in multiple ejections, most of the pulsations do not result in mass ejections.

Whether a particular assumed rate and distribution of heating leads to ejections or not is not a monotonic function of the heating rate. A potential interpretation of

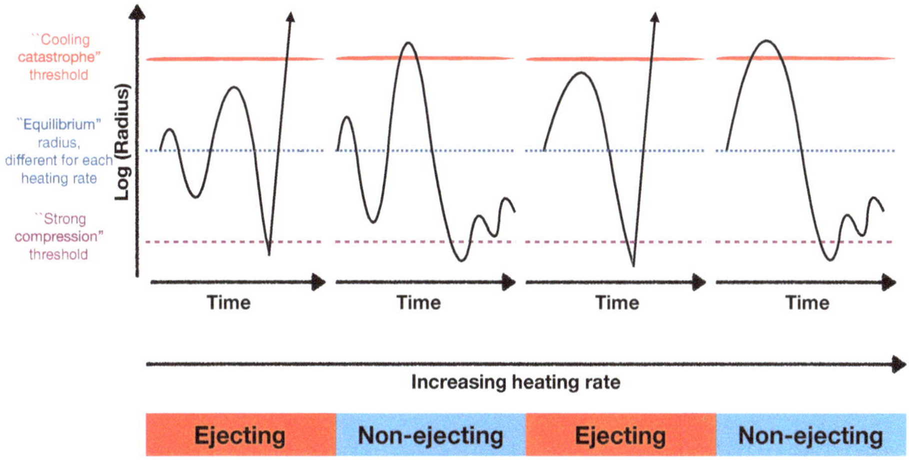

Figure 8.5. A schematic figure showing the hypothesized, simplified model of the envelope behavior seen in one-dimensional simulations performed by Clayton et al. (2017). Four examples of evolution of the radius with time correspond to four different heating rates, which increase to the right. Upper and lower radius thresholds are shown, representing radii that must be achieved for a model to undergo catastrophic cooling and mass ejections, respectively. In order for a star to eject mass, it must reach the lower threshold before the upper one. This gives rise to a complex structure of ejecting and non-ejecting regions in the heating-rate parameter space.

this is presented in Figure 8.5. Ejections can take place once the amplitude of pulsations becomes large enough that the compressional amplitude exceeds a "strong compression" threshold, the condition for which is not well-established but probably corresponds to a minimum radius or maximum speed of contraction. For expanding phases of the pulsation, the envelope can pass a "cooling catastrophe" threshold, beyond which it becomes optically thin and experiences catastrophic cooling. If the envelope reaches the cooling catastrophe threshold before the strong compression threshold, energy losses will prevent envelope ejection. However, a further increase of the heating luminosity may lead to the retention of enough energy to allow for ejections to resume. Additional increase in the heating luminosity may cause the cooling catastrophe threshold to be reached during the first pulsation of each growth cycle, causing mass ejection to fail.

Overall, Clayton et al. (2017) suggested that the interplay between the timescale of pulsation growth and the pulsation period determines the success of the ejections, but this interplay cannot be predicted in each case without detailed calculations. We also refer the reader to the caveats regarding one-dimensional calculations of this kind in Chapter 7, especially Section 7.2.1.

In one-dimensional simulations that used "self-consistent" time-dependent heating, where the heating rate increased with time, a somewhat different delayed dynamical instability was found (Ivanova 2002; Han et al. 2002). This instability manifested itself as pulsations which became stronger with time. It is argued that this delayed dynamical instability may lead to envelope removal either during a single pulsation if its amplitude is large enough, or also during a series of mass ejection episodes which would take place during each pulsation.

References

Clayton, M., Podsiadlowski, P., Ivanova, N., & Justham, S. 2017, MNRAS, 470, 1788
Han, Z., Podsiadlowski, P., Maxted, P. F. L., Marsh, T. R., & Ivanova, N. 2002, MNRAS, 336, 449
Ivanova, N. 2002, DPhil thesis, Balliol College, Oxford
Ivanova, N., Justham, S., Avendano Nandez, J. L., & Lombardi, J. C. 2013, Sci, 339, 433
Ivanova, N., Justham, S., & Podsiadlowski, P. 2015, MNRAS, 447, 2181
Ivanova, N., & Nandez, J. 2018, Galax, 6, 75
Ivanova, N., & Nandez, J. L. A. 2016, MNRAS, 462, 362
Ricker, P. M., & Taam, R. E. 2012, ApJ, 746, 74

Natalia Ivanova, Stephen Justham and Paul Ricker

Chapter 9

The Outcomes of CE Simulations

Let us not forget that the ultimate goal of the back-of-the-envelope calculations, of the multiple thought experiments regarding which processes are important and which are not, and of the many expensive numerical simulations, is to make sense of reality. To do this we must connect the predictions of theory to the observational data available. The harsh reality of common-envelope astrophysics is that the observational constraints on theory are extremely limited. Essentially, what we have are the orbital properties of binary stars that are thought to be post-common-envelope systems, the properties of the potential remnants of stellar mergers, and the shapes of planetary nebulae affected by common-envelope evolution (CEE). In this Chapter we discuss the predicted outcomes of successful common-envelope ejection. Additional observational features and constraints will be discussed in the next Chapter.

9.1 The Mass of the Initial and Remnant Core

In a successful common-envelope event, at some point all or part of the donor's envelope leaves the binary system. The portion of the donor that is left behind can be called a remnant or a stripped star. An important question here is the mass of this stripped star and whether one can predict it from the initial core mass, or whether the final mass also depends on other details of the star's initial structure.

In simplified treatments of common-envelope evolution, the "core mass" is often treated as a quantity that is known in advance of the common-envelope phase. In reality the distinction between the pre-common-envelope core and the envelope is not always well-defined. So we do not know in advance what the remnant core mass should be, if the envelope is successfully ejected. Hence we also do not know how much envelope mass needs to be ejected in order to form the stripped remnant. Because most of the binding energy is associated with material deep inside the donor, small differences in the final core mass can make a large difference in the

doi:10.1088/2514-3433/abb6f0ch9

energy expenditure required to remove the envelope (see Chapter 3, especially Figure 3.1).

Separately, we need to know the structure of the stripped star in order to predict the post-common-envelope binary evolution. In some cases, this may include additional mass transfer from the post-common-envelope remnant core (as described further later in this section).

There is no complete answer to this question yet.

The common, simplified approach of assuming that what remains is the initial stellar core, and what is removed is the initial stellar envelope, works well only in the case of low-mass giants with degenerate cores. The density contrast between the core and envelope in these cases is high, while the mass of the hydrogen-burning shell at the base of the envelope is negligible. It is also known that if the convective envelope of a low-mass giant is almost entirely stripped, so that the remaining hydrogen shell is below a minimum envelope mass, which is a function of the core mass, then the remaining envelope collapses onto the core, and the core becomes a white dwarf (Deinzer & von Sengbusch 1970). (Or, for somewhat fine-tuned timing of the envelope ejection, a helium-burning "hot subdwarf"—see, e.g., Han et al. 2002.)

In the case of more massive stars, with non-degenerate cores, the problem is more complicated. In these stars the mass of the thick hydrogen shell, which is between the core and the convective envelope, is non-negligible. While it is still less massive by a factor of ten or so than the envelope, the binding energy of this hydrogen shell can itself be a few to 100 times larger than the binding energy of the convective envelope. Determining which mass is to be removed by the envelope is critical for the determination of the final orbital separation. If the final orbital separation is a strong function of the energy expenditure, the orbital separation in the post-ejection binary can be different by up to two orders of magnitude, depending on the chosen definition of the core boundary—see Figure 3.1, which shows that the binding energy spans orders of magnitudes within the region where the separation line between the ejected and retained material can be assumed to be.

The range of the plausible core masses lies between a minimum value, corresponding to the hydrogen-exhausted core, and a maximum value, corresponding to the transition between the radiative zone (which includes the burning shell) and the bottom of the outer convective envelope. The transition between the radiative and convective zones, however, is also not uniquely determined in the literature. It may be the location where the entropy profile changes from an increasing slope to a flat one (Tauris & Dewi 2001), or perhaps the location where the effective polytropic index is discontinuous (Hjellming & Webbink 1987). The above description is relevant only to stars with a well-developed deep convection zone. For massive giant stars in which the outer convective zone occupies a small fraction of the total envelope mass (see discussion in Pavlovskii et al. 2017), there is no established understanding of the maximum possible core mass (other than the total mass of the star!). A further complication is the fact that the location of the boundary between the radiative and convective zones can change through the CE phase, and hence even the post-mass-transfer, pre-plunge-in structure cannot always be predictive, though it is less bad than that of an unperturbed single star.

The conditions used so far can be organized into the following three main categories.

1. Conditions based on thermodynamic quantities include:
    - where $\partial^2 \ln \rho / \partial^2 m = 0$ within the hydrogen-burning shell (Bisscheroux 1998);
    - where the function of the binding energy $y = \sinh^{-1}(E_{\rm bind})$ transitions between a sharply increasing and a fairly slowly increasing behavior in the outer envelope (Han et al. 1994);
    - where the value of P/ρ is at its maximum within the hydrogen burning shell; this could be described as the point of maximum local sonic velocity, or maximum compression point (Ivanova 2011).
2. Conditions based on chemical composition include:
    - where the enclosed mass contains less than ten percent hydrogen by mass (Dewi & Tauris 2000);
    - the location where the hydrogen mass fraction $X = 0.15$ (Xu & Li 2010).
3. Conditions based on nuclear energy generation rate include:
    - the location of the maximum specific rate of nuclear energy generation within the H shell (Tauris & Dewi 2001);
    - the location of the maximum specific rate of nuclear energy generation, combined with a condition on the mass of the remaining envelope, which itself is a function of the evolutionary status of the donor (for low-mass red giants and asymptotic giant branch stars, De Marco et al. 2011);
    - the location where the specific nuclear energy generation rate falls below some threshold (De Marco et al. 2011).

Some examples of the definitions and their effect on the inferred envelope binding energy for one star are shown in Figure 3.1. Note that the relative values of the envelope binding energy which are a consequence of these definitions is not fixed.

No single definition can be applied to every star. Some of the core definitions are based on features that are not present in all cases. For example, the condition using the mass of the remaining envelope is based on low-mass giants, but this class of criteria cannot be extended to all types of donors, nor can it be obtained easily even for the class of donors to which it applies. Moreover, some conditions are not unique. For example, the condition $\partial^2 \log \rho / \partial^2 m = 0$ does not always give a unique answer for massive stars. Finally, population synthesis calculations only have access to such internal structure information if they adopt full stellar structure calculations, which is currently atypical.

Some conditions might give similar results, though this does not prove they are the correct choice. For example, work covering a wide range of masses (Kruckow et al. 2016) has found that for massive stars, the core mass determined by the maximum compression point can be close to the value determined by where the enclosed mass is less than ten percent hydrogen (Dewi & Tauris 2000).

Most of the core definitions are somewhat ad hoc and may not carry much physical meaning. However, they are usually numerically well-defined and relatively easy to calculate, and they can often be used to compare outcomes from different population studies. Several groups have produced fitting formulae based on stellar models that are useful for population synthesis studies (e.g., Loveridge et al. 2011; Wang et al. 2016). Readers of studies which make use of these formulae should be careful to understand which energies are included (thermal, internal, recombination) in the binding-energy fits they use, and which core definition is used. More subtly, these energetic fitting formulae will typically not be self-consistent with the semi-analytic fitting formulae and recipes used to model binary-star evolution. The core-mass definitions used in fitting formulae based on single-star evolution can be very different from those that are used to calculate binding energies by those who make fitting formulate for binding energies. Fitting formulae for core radii at the onset of CE can be found in Hall & Tout (2014).

In addition to the problem of the core mass at the time when the envelope is departing, it has become clear that one cannot neglect the response of the newly-exposed core during its thermal-timescale recovery. The character of this recovery is linked to the definition of the core. For example, Ivanova (2011) argued that if the envelope is removed down to some position above the maximum compression point, the mass lying above the maximum compression point will re-expand on the core's thermal timescale. Fast re-expansion can lead to an episode of rapid mass transfer onto the companion. For a neutron star or black hole companion this might lead as well to hypercritical accretion, or even to a "sort-of-new" common-envelope episode in which the accreted material expands to engulf the original donor. After thermal relaxation, the cooling core usually proceeds to a contracting sequence, so the mass eventually observed in a post-common envelope binary can be less than the immediate post-ejection "core" mass.

9.2 Properties of Post-common Envelope Binaries

The overall energy budget, the complete energy formalism, and the energy components have been described in Section 3.6. Here we will provide an overview of what has been learned from simulations. This overview is mainly restricted to the case of binaries with low-mass giant donors, and further only those which eject their envelopes on a dynamical timescale (specifically, the simulations in Nandez & Ivanova 2016). Any extrapolation to massive donors is not recommended, or should be done with caution.

The final energies in the context of the energy formalism:

As discussed in Chapter 3, the efficiency in the energy formalism accounts for the fraction of the orbital energy release that is used to relocate the envelope to infinity. In simulations that eject the envelope, this efficiency is ~45%–80%. The remaining energy primarily exists as kinetic and thermal energies of the escaped envelope; the envelope's self-potential energy at infinity is small. The fraction of the released orbital energy that is removed to infinity is usually larger for less massive giants. For giants of the same mass, the fraction is larger for more evolved giants; see Figure 9.1.

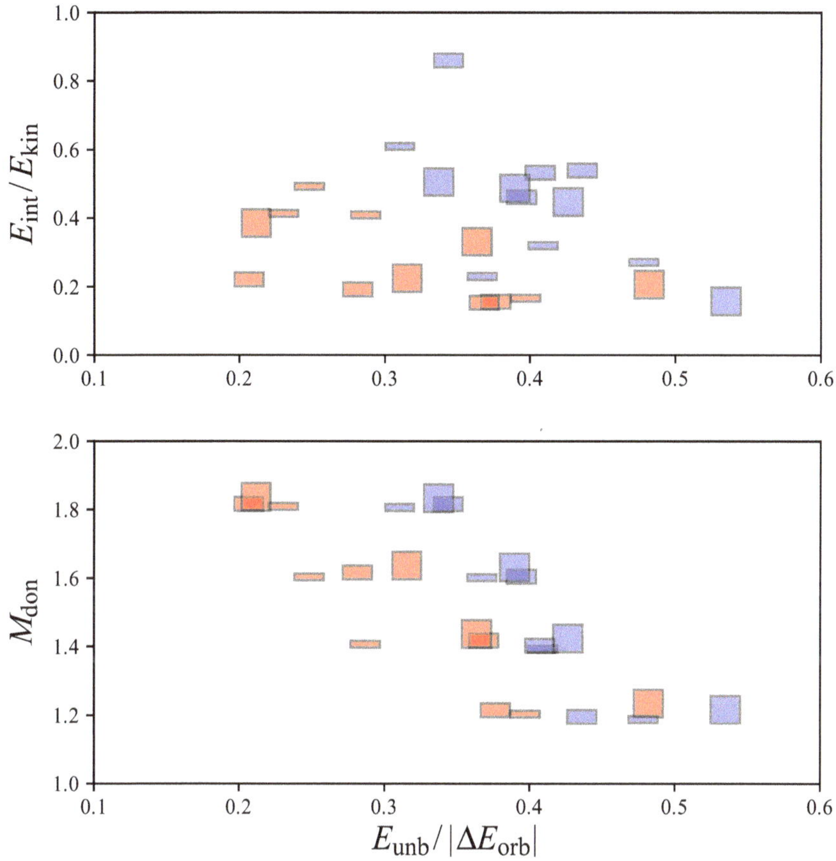

Figure 9.1. The "excess" energy in the ejected envelope at infinity. E_{unb} includes all energies that the ejected material has—internal E_{int}, kinetic E_{kin}, and potential E_{pot}, with $|E_{pot}| \ll E_{kin}$). The energy that the ejected envelope has at infinity describes "α-inefficiency": $E_{unb}/\Delta E_{orb} = 1 - \alpha_{CE}$. The aspect ratio of the symbols is related to the mass of the companion: the most narrow rectangles are for $0.32 M_{\odot}$ companions, intermediate-thick rectangles are for $0.36 M_{\odot}$ companions, and squares are for $0.42 M_{\odot}$ companions. The red color represents red giants with $0.32 M_{\odot}$ cores, and the blue color represents $0.36 M_{\odot}$ cores. All the donor stars had initial masses between $\approx$ 1.2 and 1.8 $M_{\odot}$. This plot uses the results of three-dimensional simulations published in Nandez & Ivanova (2016).

The ratio of the energy stored as thermal energy to the kinetic energy increases for more evolved giants. It must be noted that in these three-dimensional simulations there were no radiative losses.

Center of mass velocities:

The energy of motion of the center of mass of the post-CE binary usually does not play a crucial role in the energy budget, and hence is not decisive for predicting the final orbital separation from an energy balance argument. In most simulations which analyzed the center of mass motion, this velocity was found to be of the order of a few kilometers per second (Sandquist et al. 1998; Chamandy et al. 2019). This "kick" velocity may be related to the asymmetry of the ejection during the initial stages; for example, during the early stages the matter can depart the binary with a "jet-like" morphology (Ivanova & Nandez 2018).

Eccentricities:

In simulations that do not result in merger, the final eccentricities are very small, between 0.002 and 0.05, with no noticeable trend with initial binary properties (Nandez & Ivanova 2016). A higher eccentricity value can be expected if the complete envelope ejection occurs during the dynamical plunge-in, but such ejection has not yet been observed in 3D simulations.

9.3 Characteristics of Outflows

What is the structure of the envelope material ejected in a common-envelope event? We may characterize the outflow in terms of its radial distribution and its polar and azimuthal structure. Additionally, the structure of the outflow changes with time (see Chapter 8). Existing 3D simulations necessarily cover a small part of the overall ejection, but based on the existing work we can draw a few conclusions about the outflow structure.

The low-mass initial ejecta (see more details in Section 8.1) that carry away a significant amount of angular momentum are very equatorially focused. However, some initial outflows due to early mass loss from the system may remain bound to the binary and form a circumbinary torus (MacLeod et al. 2018).

During the plunge (see more details in Section 8.2), the ejecta continue to be asymmetric. Material that is ejected earlier tends to be more concentrated along the equatorial plane, with little material ejected in the polar directions (see Figures 9.2 and 9.3).

In all of the few simulations in which the envelope removal is completed via recombination outflows (see more details in Section 8.3), the ejecta becomes more isotropic at later stages (Ivanova & Nandez 2016). An example of the spherically-averaged density and velocity profiles of the ejecta is shown in Figure 8.3.

The shapes seen in visualizations of simulations (e.g., the degree of asymmetry, whether there are "spiral-like" or "jet-like" features, and in what plane) are somewhat subjective and can be misleading. It matters how the authors chose to show the density and velocity distributions in their simulations, in the sense of what range of values they decided to use to visualize their simulations, and at what time after a plunge the shape is demonstrated (compare the top plot in Figures 9.2 and 9.4). Some features can appear only if a wide range of values (minimum to maximum) is shown, while the ejected envelope appears more symmetric if one plots using higher values of density only. Some features can appear or disappear with time; for example, some initial asymmetry may be smoothed out if the matter in a feature is followed by material with higher velocity. Also, some features are "frozen" into the ejected matter forever, such as asymmetric features in the initially ejected low-mass, high-velocity matter.

Density is an easy quantity to report from simulations; however, it is not easily obtained directly from observations. The mass of optically thick material (as, for example, can be expected for equatorial outflows) is difficult to measure, not to

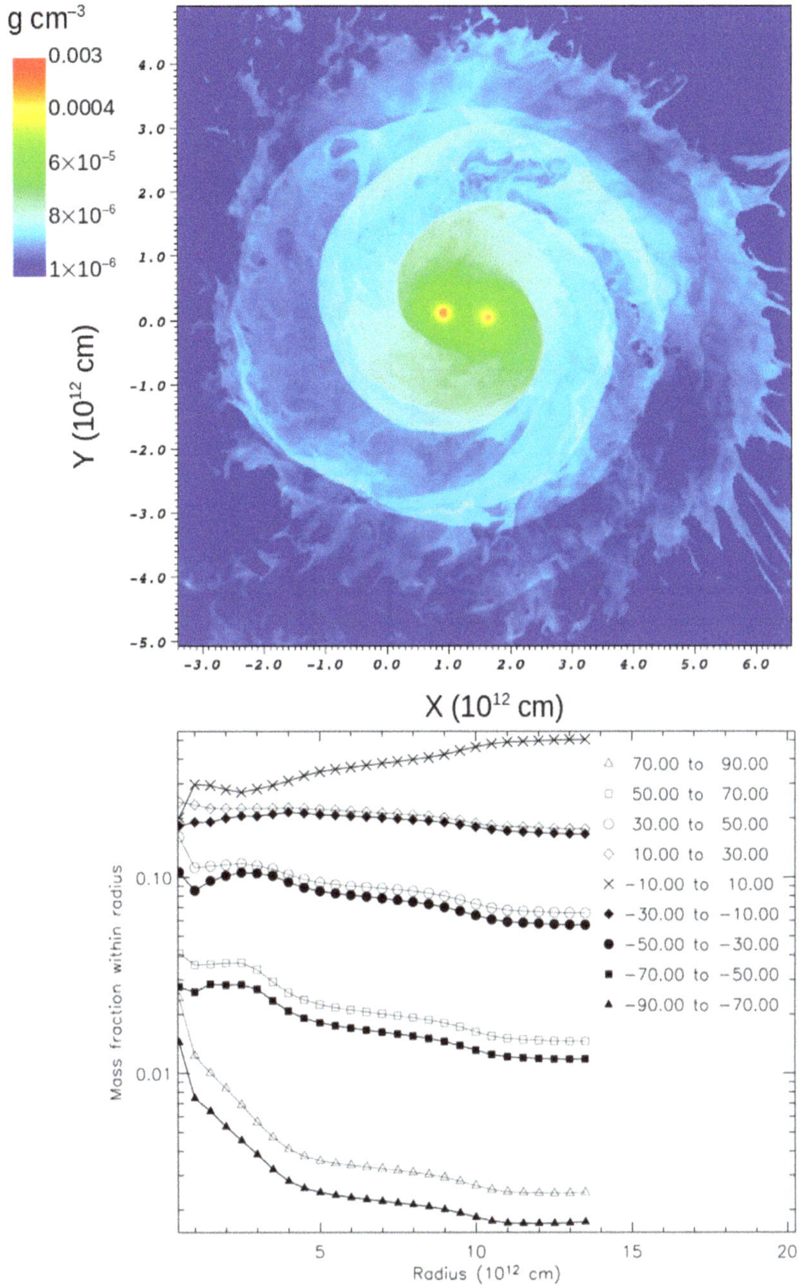

Figure 9.2. The hydrodynamic evolution of the common-envelope (CE) phase of a low-mass binary composed of a $1.05M_{\odot}$ red giant and a $0.6M_{\odot}$ companion showed at the same moment of time, 56.7 days since the start of the simulation. The top figure shows gas density in the orbital plane. The bottom figure shows enclosed mass versus radius for gas lying within different polar angle bins, as measured using the point (1.5×10^{12} cm, 0, 0) as the origin and the Cartesian z-axis as the polar z-axis. The figures are reproduced from Ricker & Taam (2012).

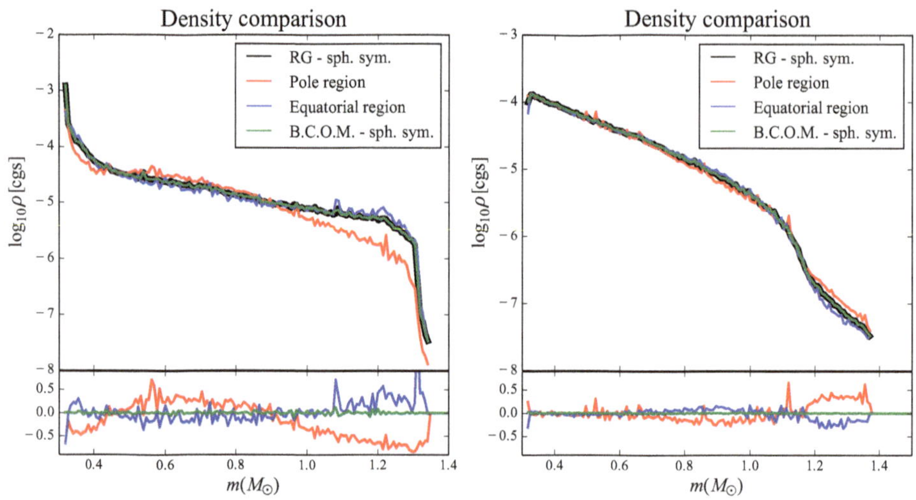

Figure 9.3. Density distributions during simulations of two different common-envelope events for a binary in which the pre-CE giant star was $1.8M_\odot$ red giant with a radius of $16R_\odot$. The pre-CE companion star was $0.36M_\odot$ for the simulation shown in the left panel and $0.15M_\odot$ for the right panel. The left panel shows the common-envelope event just before the end of the plunge-in, when outflows are significant. The right panel shows the common-envelope event at the start of the slow spiral-in, at which time there are almost no outflows. In neither case is the envelope fully ejected (with the $0.36M_\odot$ companion it will later become unbound). This is presented as a comparison of the ways to average quantities. Shown are four ways to average the density: taking the red giant core as the center and spherically averaging over all directions ($\rho_{\rm RG}$, thick black); using the same center of symmetry but only the polar region within 25° around the zenith direction ($\rho_{\rm pole}$, red); using the same center of symmetry and only the equatorial region within $\pm 15°$ around the equatorial plane ($\rho_{\rm equat}$, blue); and finally taking the center to be the center of mass of the binary, and spherically averaging over all directions ($\rho_{\rm BCOM}$, green). The bottom panels show the residuals of the three latter quantities, normalized to $\rho_{\rm RG}$, i.e., $(\rho_{\rm pole} - \rho_{\rm RG})/\rho_{\rm RG}$ (red), $(\rho_{\rm equat} - \rho_{\rm RG})/\rho_{\rm RG}$ (blue) and $(\rho_{\rm BCOM} - \rho_{\rm RG})/\rho_{\rm RG}$ (green). The figure is reproduced from Ivanova & Nandez (2016).

mention the fact that it can be hard to identify a central binary behind it. Luminosity in a particular waveband is the preferred observational quantity. However, delivering an "observable luminosity" from simulations is difficult, not only because it requires integrating along a line of sight, but also because to obtain the proper temperature one needs to include radiative losses from the ejected material into one's calculations. As material cools, in principle one would also need to model dust formation.

It should be noted that, so far, three-dimensional simulations of common envelope events have not produced polar outflows that would be substantially faster (about an order of magnitude) than the equatorial outflows during a CEE, though these faster polar outflows have been observed (see Section 10.3). Such outflows are likely to be created during a post-CE phase, for example due to a fast wind from the remnant central object (see, e.g., a simulation that includes such a wind García-Segura et al. 2018). It is not fully excluded that some polar outflows could be created by a jet during one of the phases of a common-envelope event, but jets themselves have never been formed self-consistently within common-envelope simulations

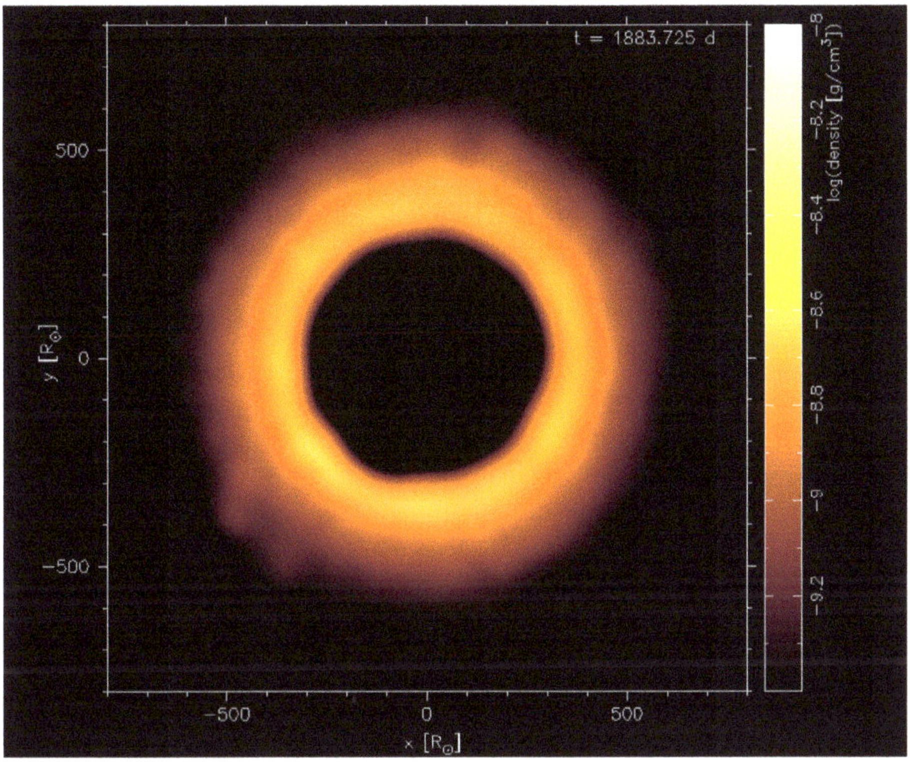

Figure 9.4. A post-CE "nebula" from a three-dimensional simulation between a $1.6M_{\odot}$ red giant (with a $0.32M_{\odot}$ core) and a $0.36M_{\odot}$ white dwarf companion. The simulation is shown about 1000 days after the start of the plunge. This is the case of a complete CE ejection. The figure is reproduced from Ivanova & Nandez (2018).

(although nearly polar-direction ejections have sometimes been observed in simulations, with no clear evidence of whether they are physical features or numerical artefacts).

9.4 Can Angular Momentum Conservation Be Used to Predict CEE Outcomes?

Application of energy conservation to common-envelope evolution is a mess, no doubt. As was discussed in Chapter 3, energy takes a number of forms in common-envelope systems, and it can be lost via radiation. On the other hand, angular momentum takes fewer forms, and it is not lost to radiation in any significant way. To complicate matters further, kinetic and heat energy can even be generated from nuclear reservoirs, which are not regularly taken into account. Unlike energy, bulk angular momentum is unlikely to be created in significant amounts from microscopic motions. This is why angular momentum conservation is normally used to consider the orbital evolution before and during stable RLOF, and during mass loss from binaries. It seems to be perfect.

The beauty of angular momentum conservation has led in the past to the proposal that it can be used to predict the outcome of *unstable* mass transfer, assuming that any angular momentum loss is the orbital angular momentum times some multiplier (Nelemans et al. 2000; Nelemans & Tout 2005). This has become known as the γ formalism. In this formalism, the outcome is predicted very similarly to the orbital evolution of a non-CE binary during mass loss, but using γ_{CE} values inferred from observations of double white dwarf systems. It is also important to remember that originally this formalism was created to explain systems that do not proceed through a significant spiral-in.

There is always a catch in the apparent simplicity. Here the catch is that the final orbital angular momentum of the post-CEE binary is a tiny fraction of the initial orbital angular momentum, and hence the values of γ_{CE} must be known to very high precision (one percent or less) in order to have useful predictive power for post-CE orbital separations. A small variation can change the outcome wildly, from a merger to orbital expansion (see Webbink 2008 and Woods et al. 2010; for a more formal mathematical explanation see Woods et al. 2012; and for a detailed review of this formalism see Ivanova et al. 2013).

As was discussed in Ivanova et al. (2013); angular momentum is not expected to be a dominant factor in determining the final state of any CEE where spiral-in is significant, because most loss of total angular momentum is expected to take place at wide separations. Indeed, as was demonstrated in Ivanova & Nandez (2016), most of the angular momentum is lost from the binary *before* the plunge-in, during L_2/L_3 outflows.

We can appreciate the complexity of the task of obtaining an accurate γ_{CE} from numerical studies by considering the simulations with L_2 outflows performed by Chen et al. (2018): before the plunge, γ for outflows varies greatly during the simulation of an individual system and between simulations of different systems. No unique value can explain all the systems. One also should recall that many three-dimensional hydrodynamics codes cannot conserve angular momentum to the required precision; see Chapter 4.

With the coming era of large surveys, the possibility to derive γ_{CE} values from observed post-CE binaries is not completely ruled out. However, these values should not be inferred from systems having different ranges of post-CE masses and post-CE periods. Unfortunately, an accurate extrapolation from one type of post-CE systems to another type of post-CE systems is no more theoretically expected for the γ-formalism than for the α-formalism. Thus the γ-formalism does not provide any particular benefit over the α-formalism in mapping pre-CE systems to post-CE systems.

References

Bisscheroux, B. 1998, M.Sc. thesis, Univ. Amsterdam

Chamandy, L., Tu, Y., Blackman, E. G., et al. 2019, MNRAS, 486, 1070

Chen, Z., Blackman, E. G., Nordhaus, J., Frank, A., & Carroll-Nellenback, J. 2018, MNRAS, 473, 747

De Marco, O., Passy, J.-C., Moe, M., et al. 2011, MNRAS, 411, 2277
Deinzer, W., & von Sengbusch, K. 1970, ApJ, 160, 671
Dewi, J. D. M., & Tauris, T. M. 2000, A&A, 360, 1043
García-Segura, G., Ricker, P. M., & Taam, R. E. 2018, ApJ, 860, 19
Hall, P. D., & Tout, C. A. 2014, MNRAS, 444, 3209
Han, Z., Podsiadlowski, P., & Eggleton, P. P. 1994, MNRAS, 270, 121
Han, Z., Podsiadlowski, P., Maxted, P. F. L., Marsh, T. R., & Ivanova, N. 2002, MNRAS, 336, 449
Hjellming, M. S., & Webbink, R. F. 1987, ApJ, 318, 794
Ivanova, N. 2011, ApJ, 730, 76
Ivanova, N., Justham, S., Chen, X., et al. 2013, A&ARv, 21, 59
Ivanova, N., & Nandez, J. 2018, Galax, 6, 75
Ivanova, N., & Nandez, J. L. A. 2016, MNRAS, 462, 362
Kruckow, M. U., Tauris, T. M., Langer, N., et al. 2016, A&A, 596, A58
Loveridge, A. J., van der Sluys, M. V., & Kalogera, V. 2011, ApJ, 743, 49
MacLeod, M., Ostriker, E. C., & Stone, J. M. 2018, ApJ, 868, 136
Nandez, J. L. A., & Ivanova, N. 2016, MNRAS, 460, 3992
Nelemans, G., & Tout, C. A. 2005, MNRAS, 356, 753
Nelemans, G., Verbunt, F., Yungelson, L. R., & Portegies Zwart, S. F. 2000, A&A, 360, 1011
Pavlovskii, K., Ivanova, N., Belczynski, K., & Van, K. X. 2017, MNRAS, 465, 2092
Ricker, P. M., & Taam, R. E. 2012, ApJ, 746, 74
Sandquist, E. L., Taam, R. E., Chen, X., Bodenheimer, P., & Burkert, A. 1998, ApJ, 500, 909
Tauris, T. M., & Dewi, J. D. M. 2001, A&A, 369, 170
Wang, C., Jia, K., & Li, X.-D. 2016, RAA, 16, 126
Webbink, R. F. 2008, in Astrophysics and Space Science Library, Short-Period Binary Stars: Observations, Analyses, and Results, ed. E. F. Milone, D. A. Leahy, & D. W. Hobill (Berlin: Springer), 233
Woods, T. E., Ivanova, N., van der Sluys, M., & Chaichenets, S. 2010, in AIP Conf. Ser. 1314, International Conference on Binaries, ed. V. Kologera, & M. van der Sluys (Melville, NY: AIP), 24
Woods, T. E., Ivanova, N., van der Sluys, M. V., & Chaichenets, S. 2012, ApJ, 744, 12
Xu, X.-J., & Li, X.-D. 2010, ApJ, 716, 114

Natalia Ivanova, Stephen Justham and Paul Ricker

Chapter 10

Linking with Observations

This book has mostly tried to approach the common-envelope problem from first principles, by using analytic and numerical modeling, and by considering which simulations provide appropriate approximations to the situation. This Chapter summarizes the main observational avenues that have the potential to constrain the physics of the common-envelope problem.

Attempts at constraining the physics of common-envelope evolution (CEE) using observed systems are sometimes clearly framed as searching for clues in the phenomenology, e.g., looking for systematic trends in the properties of a post-CE population. They can also be focused on specific physics questions, e.g., investigating at what velocities common-envelope material is ejected from different systems. Attempts to infer parameters such as "the common-envelope efficiency" $\alpha_{\rm CE}$ belong in neither of these categories; as described in Chapter 3, this parameterization is only indirectly related to the physical problem, and there is no reason to expect that the $\alpha_{\rm CE}$ parameter is a universal constant. However, fitting observed systems using such parameterizations remains common, and can be physically informative if interpreted carefully.

10.1 Overview

Any scientific theory or model should be tested against observational data. This is generally considered to be a minimal requirement of the scientific process. However, comparing theory and reality is not trivial when dealing with any process that happens over astrophysical timescales. For example, it would be optimistic to write an observing proposal to wait for many tens of years to watch an unstable mass-transfer episode in a binary-star system turn into a common-envelope inspiral phase and then (potentially) lead to ejection of the envelope, eventually revealing the properties of a post-common-envelope binary. That would be true even if we had guaranteed nearby examples to observe, and even if we were sure that "tens of years" is an upper limit to the likely timescale involved for those examples. In order

doi:10.1088/2514-3433/abb6f0ch10

to produce quantitative conclusions about the physics, observational evidence about common-envelope evolution (CEE) is interpreted through modeling.

This Chapter is an attempt, from the point of view of three theorists, to discuss how observations may constrain the theory. We focus on: post-common-envelope ejection binary systems (Section 10.2); nebulae around post-common-envelope binaries (Section 10.3); post-merger stars and the nebulae associated with them (Section 10.4); transients associated with common-envelope ejection or mergers (Section 10.5); and, potentially, the appearance of stars undergoing a common-envelope phase (Section 10.6). Of course, other types of observational evidence beyond those we discuss in this Chapter are potentially relevant to common-envelope physics, since stellar interactions are important in the evolution of a wide variety of stellar systems. Clearly, for example, there is some information from post-stable-mass-transfer binary systems in constraining mass-transfer stability.[1]

Broadly speaking, comparisons between observations and theoretical predictions may be performed for individual examples or for populations.

Comparisons with well-studied individual examples seem likely to be more promising for producing detailed physical understanding. For example, given systems for which we can confidently infer past system properties (e.g., a pre-CE state), we can perform before-and-after comparisons to try to constrain models for time evolution. Or when we are fortunate to have a well-studied stellar-merger transient, with well-characterized pre-merger properties (such as V1309 Sco; Tylenda et al. 2011), we may use those pre-merger properties to simulate the onset of the merger and the transient (Ivanova et al. 2013; Nandez et al. 2014).

In principle, comparisons between population models and population observations help us to make use of the information from the sky to constrain processes occurring on evolutionary timescales. On the other hand, the practice of population modeling typically introduces additional uncertainties (see Section 10.2.2 for a brief discussion), or at least other explicit and implicit assumptions beyond CE physics. Nonetheless, systematic and synoptic population comparisons should help to restrain theorists from fine-tuning model physics for individual cases: it is inconsistent to re-use the same set of initial conditions to explain two different, mutually exclusive, observed populations by changing physical assumptions. Only one set of physics can be correct.

10.2 Post-common-envelope Binary Properties

It feels almost tautologous to write that the most natural way to study the outcomes of common-envelope ejection is to find and characterize binaries that have been through CEE.

In the early years after the mechanism had been proposed, simply identifying any probable post-CE close binaries was a paramount goal. The abstract of Paczynski (1976) includes the statement that "discovery of a short-period binary being a

[1] We can also make inferences from systems during stable mass transfer. Although, as described in Chapter 5, mass transfer may become unstable after an extended phase of apparently stable RLOF.

nucleus of a planetary nebula would provide very important support for the evolutionary scenario presented in this paper." The "short-period" in that statement is crucial; as Paczynski noted, some planetary nebulae were already thought to contain binary stars (e.g., Kohoutek 1967). Simply identifying a binary star in a planetary nebula would not be evidence that the material in the nebula was ejected through CEE, e.g., if the binary was wide enough for one of the stars to have ejected its own envelope as an AGB star. However, Bond (1976) identified the central star of the planetary nebula Abell 63 with the known variable UU Sge (following indications of variability suggested by Hoffleit 1932 & Abell 1966). Then Miller et al. (1976) and Bond et al. (1978) established its orbital period to be approximately 11 hr, with two low-mass stars—one hot, O-type, star and one cool, K-type, star—separated by roughly 3 $R_\odot$, i.e., far too close to contain an AGB star. Bond et al. (1978) discuss the origin of the configuration, and state that Paczynski (1976) provided "an extraordinarily accurate description of the UU Sge/Abell 63 system," with the hot star that has ionized the nebula presumed to be the core of a giant star that has been exposed by common-envelope ejection, and the current close separation a consequence of orbital decay during CEE.

So, soon after the paper regarded as central for developing common-envelope evolution as an important process in the formation of close binary systems (Paczynski 1976), observers began to find binaries which fit the expected post-CE properties, and which were surrounded by material consistent with being the ejected envelope. Post-CE close binary central stars of planetary nebulae (close binary CSPNe) are examples of post-CE configurations relatively soon after the ejection of the envelope. For recent reviews see, e.g., Jones (2017) or Jones (2018). In Section 10.3 we will look at how the nebulae themselves may provide constraints.

However, the description "post-common-envelope binaries" provides yet another case in which astronomical terminology can be unhelpful. The term "post-common-envelope binaries" (PCEBs) is often taken to refer to only a subset of the actual post-common-envelope binaries. Specifically, it is commonly used to refer to systems which resemble UU Sge, the binary CSPN of Abell 63, in containing one hot component (typically a white dwarf [WD] or proto-WD) and one main-sequence star. Systems called PCEBs are detached, with orbits close enough for a prior phase of common-envelope evolution to be suspected, if not always guaranteed. Strikingly, even close binaries containing a hot subdwarf and a main-sequence star—as in UU Sge—are not always included as PCEBs, despite the fact that subdwarfs indicate a higher probability of being post-common-envelope binaries. Here we do not follow this restrictive convention for the PCEB term, and aim to specify more clearly what type of binary we mean, but we encourage readers of this book to be careful when reading the broader literature.

10.2.1 Inferred Constraints from Individual Post-CE Systems

Given a binary for which the most recent orbital transformation is thought to have been as result of common-envelope interaction, we may be able to infer the pre-CE properties. This case can be reasonably made if we think the pre-CE donor star radius can be inferred from its core mass; such a relationship exists for hydrogen-shell-burning giants with degenerate cores (combining the core-mass–luminosity relation of, e.g., Refsdal & Weigert 1970; Paczyński 1970; Kippenhahn 1981 with the Hayashi line), and we take the mass of the pre-CE core from the mass of the post-CE exposed core (see Chapter 9 for caveats). The apparent uncertainty in the pre-CE state decreases if we can infer the total mass of the pre-CE donor star via, e.g., inferring the overall age of the system, and combining that with the inferred pre-CE stellar radius to infer the initial total mass of the pre-CE donor star.

We cannot provide a comprehensive review, but we highlight some ways in which inferences from observations have been used to try to draw conclusions about CE physics.

10.2.1.1 Inferences from Systems with Wide Post-CE Separations

If we analyze CEE in a simple energy-balance picture then, for a given pre-CE binary configuration, a wider post-envelope ejection binary means the ejection was "more efficient," i.e., less orbital energy release from inspiral was needed to eject the envelope. One common line of investigation has been to study unusually wide post-CE binaries, and so try to infer either a maximum "efficiency" for the energetics of that CE ejection, or to conclude that recombination energy must have been used in ejecting the envelope in that case.

Even a relatively small amount of orbital energy release may be able to eject the envelope, if the envelope is loosely bound. This situation is not, in principle, troublesome. We remind the reader that large, cool envelopes can be formally energetically *unbound* when the recombination energy reservoir is taken into account (e.g., Paczyński 1968; Han et al. 1994), i.e., in principle they have no binding energy if recombination can be suitably triggered. So in those cases, wide post-CE binaries could easily be consistent with a simple energy budget that only includes the envelope binding energy and orbital energy release. However, this is only in the sense that there was enough energy released, but not in a sense that the final separation can be predicted by the initial energy balance (see also all the caveats and discussion in Chapter 3).

Nonetheless, it would be helpful if inferences from observations can quantify the extent to which recombination energy can be used, in which cases, and how much shrinkage must happen in order to trigger recombination-driven outflows.

Moreover, in trying to understand post-CE systems, we should not only focus on the energy budget. The main application of CE evolution is to predict where the inspiral stops. Would two different inspirals with identical orbital energy input lead to identical outcomes if one involves a massive companion that hardly changes separation, and the other involves a low-mass companion that plunges deep into the envelope? If we take it as given that post-CE separations tell us the orbital separations at which the companion no longer experienced orbital drag from the envelope, then wide post-CE separations indicate systems for which the envelope was ejected with relatively wide separations. Perhaps the pre-CE envelopes in wide post-CE cases were so loosely-bound as to be ejected unusually promptly?

A small number of systems have been studied as suspected wide post-CE systems. In particular, because of its moderately-wide 22 day orbit, IK Peg is regarded as an example of evidence that CE ejection may be extremely efficient or involve an additional energy sources, or that the effective binding energy of the envelope in the pre-CE system was low (e.g., Davis et al. 2010; Zorotovic et al. 2010; Davis et al. 2012; Rebassa-Mansergas et al. 2012; Iaconi 2019; we remind the reader that recombination energy is sometimes thought of as an "extra" energy source, and is sometimes included in the calculation of the envelope binding energy).[2] Similarly, Maxted et al. (2002) suggest that the properties of PG 1115+166, a double white dwarf system with a 30 day orbital period and in which both components are 0.7 $M_{\odot}$, are consistent with ejection of an envelope with help from recombination energy. Kruse & Agol (2014) discovered a wonderful—self-lensing—binary with an 88 day orbital period, which Zorotovic et al. (2014) concluded was only the second known system, after IK Peg, "that clearly requires an extra source of energy, beyond that of orbital energy, to contribute to the CE ejection," for which the natural candidate is recombination energy. Webbink (2008) also discusses in this context the prior evolution of two wide systems containing *accreting* WDs, i.e., systems often excluded from discussions of post-CE properties, the recurrent novae and potential SN Ia progenitors T CrB and RS Oph. These have even longer present-day orbital periods than IK Peg or KOI-3278, over 220 days for T CrB and over 450 days for RS Oph. Webbink (2008) concludes that RS Oph and T CrB are evidence that recombination energy can be used in envelope ejection. We repeat that, if these are post-CE systems, they would also indicate that CE ejection, and termination of the inspiral, can occur after relatively shallow inspiral.

All of IK Peg, KOI-3278, T CrB, and RS Oph contain relatively massive companion stars, as compared to the broader sample of known post-CE binaries containing a white dwarf and a non-degenerate companion. One potential interpretation of their wide separations for CE physics has been that more massive secondary stars may be linked with more efficient CE ejection (e.g., Zorotovic et al. 2014; Iaconi 2019). Alternatively, or complementary to that, the fact that three of them contain WDs more massive than the Sun is consistent with the pre-CE donor stars in those cases having very loosely-bound AGB envelopes (i.e., if recombination energy is included in the binding-energy calculation; Webbink 2008), so they could reasonably be interpreted as evidence in favor of the importance of recombination energy in some CE ejections.

However, we caution that wider supposed post-CE binaries may be explained through alternative formation scenarios, for which there is considerable theoretical freedom—notably in both the conservativeness and stability of mass transfer. Systems containing low-mass white dwarfs may be naturally explained through stable mass transfer (e.g., the 40 day orbital period for Regulus; cf Chapter 1 and Figure 1.3). However, it seems less widely appreciated that white dwarf masses that

[2] Davis et al. (2012) explicitly note that the pre-CE envelope in IK Peg may have been formally unbound, when including the thermal and recombination-energy terms in the binding energy, although they speculate that in this case the system may have "avoided the CE phase" while ejecting the envelope.

would normally be associated with asymptotic giant branch stars can be produced from intermediate-mass stars that are stripped into non-degenerate, helium-rich stars. van der Sluys et al. (2006) discuss this channel in the context of PG 1115+166. In particular, van der Sluys et al. (2006) note that exposed helium cores above approximately $0.8M_\odot$ expand significantly as giants, which might lead to a post-CE mass-transfer phase from the helium giant. (Maxted et al. 2002 also discuss whether the present-day orbit of PG 1115+166 may have widened since the previous CE phase, in their case speculating on transfer during a "born-again" red giant phase during the cooling of one of the white dwarfs, but also suggest that recombination energy may have been involved assisting the CE ejection.)

The population calculations of Davis et al. (2010) are a commendable example of a population study that was careful to consider whether potential post-CE binaries might have instead formed through stable mass transfer. Davis et al. (2010) could comfortably explain some wide potential post-CE binaries with low-mass white dwarfs or hot subdwarfs (FF Aqr, V651 Mon, V1379 Aql) through stable mass transfer (see also, e.g., Webbink 1979; de Kool & Ritter 1993; Iaconi 2019). By contrast, with their assumptions, Davis et al. (2010) cannot reproduce the population of relatively-wide (10–100 days) binaries containing a white dwarf more massive than the Sun and a non-degenerate companion less massive than a few times that of the Sun, such as IK Peg. However, as noted in Chapter 5, population calculations have generally tended to underestimate mass transfer stability as compared to our best present-day understanding. Therefore, it may be possible that a route exists to produce more of these moderately wide, supposedly post-CE systems via stable mass transfer instead of CEE.

10.2.1.2 Lowest-mass Companions for CE Ejection

A different type of extreme bound on CE physics is provided by searching for the minimum companion mass that can successfully eject a stellar envelope. There is nothing immediately problematic about the energy budget for ejections by extremely low-mass companions, similar to the discussion regarding long-period post-CE binaries in the previous subsection (see also Chapter 3). Nonetheless, these extreme examples of apparent CE ejection are fascinating.

Strong observational evidence that substellar companions can at least survive inspiral through a giant envelope was provided by Maxted et al. 2006; who found a 0.05 $M_\odot$ brown dwarf companion in a 116 min orbit with a white dwarf. The fact that the white dwarf is only $0.4M_\odot$ is consistent with the evolution of a first giant-branch star having been terminated by common-envelope ejection involving that brown dwarf companion. The brown dwarf companion to the white dwarf in GD1400 (Farihi & Christopher 2004) was also later found to be in a close orbit (approximately 10 hr; Burleigh et al. 2011), consistent with being a post-CE system. Another recent example is EPIC212235321, a post-CE system containing a brown dwarf with an orbital period less than 70 min (Casewell et al. 2018).

Close substellar companions have also now been identified around several hot subdwarf stars (Geier et al. 2011; Schaffenroth et al. 2014, 2015, 2019). As this book was being finalized, Kramer et al. (2020) posted a preprint showing their simulations modeling the formation of hot subdwarfs through common-envelope ejection by substellar and very-low-mass companions. They argue that their simulations find a companion mass limit for envelope ejection which is consistent with the observed systems, concluding that companions of masses down to 0.03 $M_{\odot}$ are able to trigger "a dynamical response of the stellar envelope."

10.2.1.3 Broader Inferences from Post-common-envelope Binaries

Above we have described how extreme examples from the observational sample of post-CE binaries might be used to investigate CE physics. Similar investigations can be performed for the whole sample of post-CE binaries, typically by fitting at least the $\alpha_{\rm CE}$ parameter (see Chapter 3; note especially that the commonly-used simple parameterizations of CE energetics are probably, at best, incomplete descriptions of the physics).

Nelemans et al. (2000) inferred the combination of parameters $\alpha_{\rm CE}\lambda$ during the final common-envelope ejection from a sample of 11 double white dwarf systems, finding some systems with an inferred range of $\alpha_{\rm CE}\lambda$ above 1. Only for WD 0957-666 did the inferred range of $\alpha_{\rm CE}\lambda$ values not extend above 1. Nelemans & Tout (2005) performed a similar exercise for a larger sample.[3] No single value of $\alpha_{\rm CE}\lambda$ was consistent with all of their reconstructions, which is a priori unsurprising.

Later work looked for systematic patterns in the reconstructions of the common-envelope parameters. Zorotovic et al. (2010); De Marco et al. (2011); and Davis et al. (2012) all used model values for λ in performing similar reconstructions to try to break the $\alpha_{\rm CE}\lambda$ degeneracy. A selection of their results is shown in Figure 10.1.

Davis et al. (2012) consider both the cases when only the gravitational binding energy is included in calculating λ (i.e., $\lambda_{\rm g}$) and when all the internal and recombination-energy terms are also included in λ (i.e., $\lambda_{\rm b}$), and then reconstruct $\alpha_{\rm CE}$ for a sample of post-CE binaries containing a white dwarf and a main-sequence star. For both choices of λ, Davis et al. (2012) find correlations between $\alpha_{\rm CE}$ and all of the white dwarf mass, the post-CE orbital period, and the companion-star mass. In Figure 10.1 we show a subset of their results when adopting $\lambda_{\rm b}$. For that assumption, Davis et al. (2012) find a statistically-significant correlation between $\alpha_{\rm CE}$ and the white dwarf mass, with $\alpha_{\rm CE}$ decreasing for more massive white dwarfs. In this case the range of $\alpha_{\rm CE}$ values they infer extends up to roughly 2, clusters between approximately 0.2 and 0.6, and decreases to as low as $\alpha_{\rm CE} \approx 0.02$ for more massive white dwarfs that were inferred to have lost very loosely-bound envelopes. Their best-fit line to the inferred $\alpha_{\rm CE}$ values when using $\lambda_{\rm b}$ decreases from $\alpha_{\rm CE} \approx 1$ to $\alpha_{\rm CE} \approx 0.06$ across the range of white dwarf masses in their sample.

Zorotovic et al. (2010) fit an observational sample of post-CE binaries using calculated $\lambda_{\rm g}$ values, but assuming that a fraction of the internal and recombination energy can be used to help eject the envelope, and that this fraction is equal to $\alpha_{\rm CE}$. They find no evidence for a systematic dependence of $\alpha_{\rm CE}$, concluding that "[if] there

[3] Nelemans & Tout (2005) also showed that trying to fit the *first* mass transfer phase with unstable mass transfer and a CE spiral-in does not work. This is now thought to be because those mass transfer phases are stable; see Chapter 5.

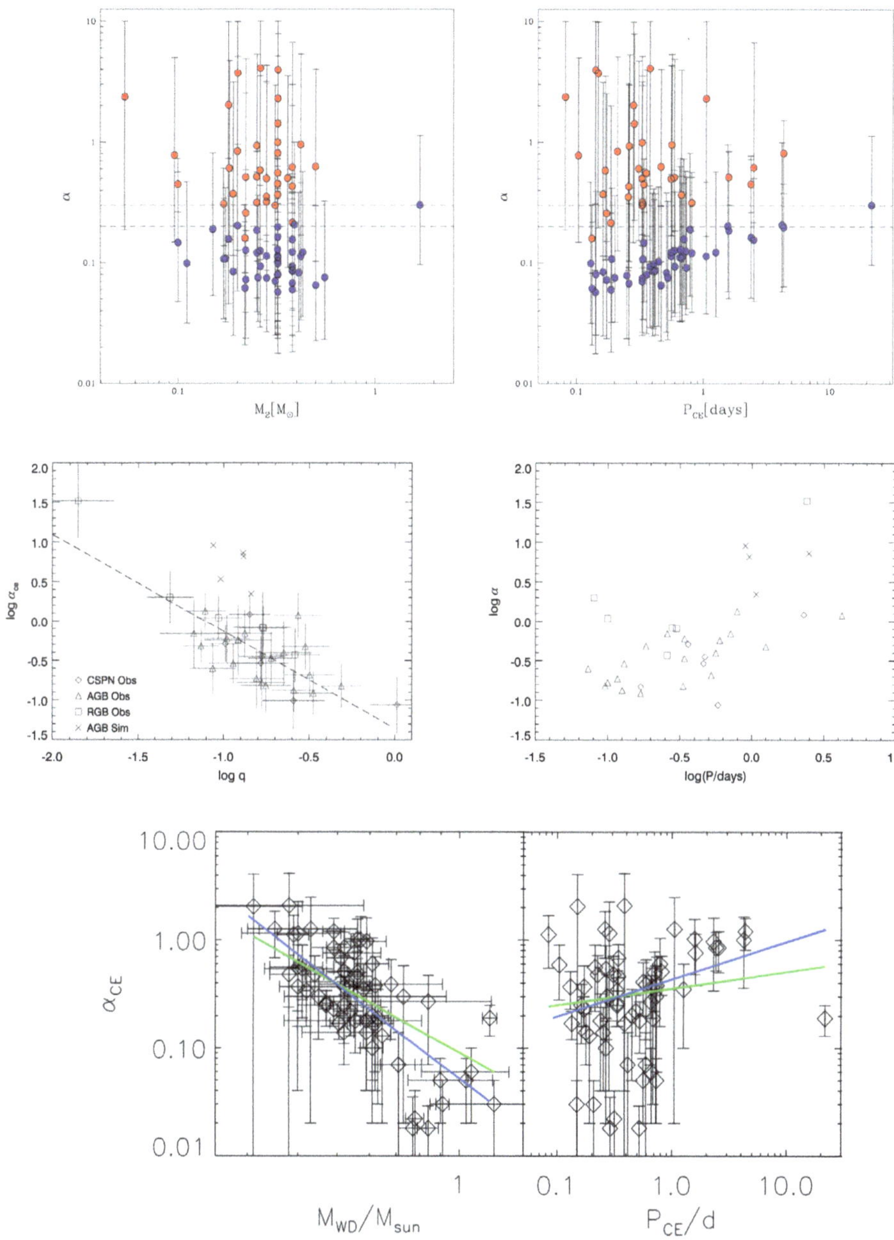

Figure 10.1. Inferred values of $\alpha_{\rm CE}$ from Zorotovic et al. (2010) (top row), De Marco et al. (2011) (middle row), Davis et al. (2012) (bottom row). In the top row, red symbols are for systems thought to have undergone CEE on the first giant branch, while blue symbols indicate post-AGB systems; in the middle panel, × symbols indicate simulations, not inferences from observed systems. In each row the right panel shows $\alpha_{\rm CE}$ against the observed post-CE orbital period. The left panels show $\alpha_{\rm CE}$ against the secondary-star mass (top row), the pre-CE mass ratio (middle row), and the post-CE white dwarf mass (bottom row). Credit: top figure: Zorotovich et al. (2010), reproduced with permission from Astronomy & Astrophysics, © ESO; middle figure: Reproduced from De Marco et al. (2011) © 2011 The Authors. CC BY; bottom figure: Reproduced from Davis et al. (2012) © 2012 The Authors. CC BY.

is a universal value of the CE efficiency," $0.2 < \alpha_{CE} < 0.3$ (see the top panels in Figure 10.1)

De Marco et al. (2011) included an approximation to the thermal energy of the envelope in their binding energy calculations,[4] but do not include the contribution of recombination. The range of α_{CE} inferred in De Marco et al. (2011); is broadly consistent with the Zorotovic et al. (2010) and Davis et al. (2012) papers; all but one outlier has an inferred α_{CE} consistent with being between 0.1 and 1.0. The strongest correlation they infer is between α_{CE} and the pre-CE mass ratio of the binary, with α_{CE} decreasing as the mass of the companion star increases toward the inferred pre-CE mass of the donor star, although the by-eye strength of that correlation appears enhanced by the two extreme cases (see the middle-left panels in Figure 10.1).

Given the uncertainties in these reconstructions, we do not conclude that any suggested correlation in α_{CE} has been robustly established from observations. Nonetheless, we encourage efforts to continue to look for systematic trends in CE outcomes, as long as the conclusions are carefully interpreted. For example, Iaconi (2019) provide a recent comparison between the properties of observed post-CE binaries and simulations. We also strongly encourage anyone whose results are sensitive to the assumption that α_{CE} is a global constant to be cautious; the observational evidence is against this.

10.2.1.4 Surface Abundances

If post-CE companion stars show anomalous surface abundances that are consistent with plausible compositions of the pre-CE donor star, this could be taken as evidence of pollution from the envelope of the donor star. This would not automatically imply that there is evidence for accretion during CE inspiral, since the accretion could occur during a phase of pre-CE mass transfer (e.g., before the Roche-lobe overflow becomes unstable), or from fallback after the majority of the envelope has been ejected, or from a phase of post-CE mass transfer from the exposed core onto the companion star. By contrast, any robust observational evidence *against* accretion onto the companion would be evidence against all those possibilities in the observed case(s).

We are aware of few claims for anomalous abundances in post-CE companion stars. Drake & Sarna (2003) argued that the K dwarf in the prototypical post-CE binary V471 Tau shows evidence for material processed by the CN-cycle, following a tentative suggestion of enhanced ^{13}CO in the same system (Dhillon & Marsh 1995). Miszalski et al. (2013) found that the non-degenerate component in the post-CE binary star in the center of the Necklace planetary nebula is carbon-rich.

10.2.2 Population Comparisons

Section 10.2.1 discussed trying to gain insight about CEE by inferring backwards in time from observations of post-CE binaries. A related approach is to do forward modeling of binary populations, and to compare those model population predictions

[4] They adopt a simple form of the virial theorem which, e.g., neglects any radiation pressure contribution.

to the observed populations. If our models adequately represent the physics, and if we know the population initial conditions, then the population predictions should match the true populations. Of course the observed populations are commonly a biased representation of the true populations, i.e., there are observational selection effects (in the context of modeling post-CE binary populations see, e.g., Pretorius et al. 2007; Toonen & Nelemans 2013; for a more statistical discussion of inferring selection effects for population analysis see, e.g., Mandel et al. 2019). Observational biases may not only affect which members of populations are more likely to be discovered, but can also affect which members of populations can more easily have their properties measured to higher precision.

In principle a vast amount of information about evolutionary processes is contained in the population data. An old example of using population information to draw evolutionary conclusions led to the knowledge that massive stars have shorter lives than less-massive stars, based on the populations in which they are found. (Some clusters contain massive stars *and* less massive stars, while others contain only the less massive stars. We conclude the first type are younger clusters, and the more massive stars have shorter lifetimes. Of course this conclusion also now easily comes out from theoretical models of stellar evolution, so theory and observation agree, but the point is that we use *population* information to support conclusions about processes happening on timescales that are too long to directly observe.) Given the increase in population observations from spectroscopic, photometric, and astrometric surveys, we should try to make the most of this observational data.

However, most commonly such work has modeled populations by starting from uniform-composition main-sequence binaries, which, for many post-CE populations, means that the population predictions depend on many aspects of non-CE physics. In some works this includes using fast but approximate semi-analytic recipes for stellar evolution and binary interactions. As noted previously, population models have also typically adopted parameterizations of CEE that have a functional form that is probably insufficient to capture CE physics. Of course all models are approximations, but the extent to which these parameterizations are quantitatively helpful approximations is not well known. Moreover, since the values of parameters such as "the common-envelope efficiency" are unlikely to be the same for all common-envelope episodes, changing global constants in models to fit population data and then drawing physical conclusions should—at best—only be done with great care. Well-intentioned comparisons to data may thereby produce parameter-fitting with unclear, or even misleading, physical meaning.

For population models, parameter-fitting might be regarded as calibration of the population models themselves against an observed population, conditional on all their other explicit and implicit assumptions. While calibrating models is preferable to not calibrating them, calibrating the parameters in a particular model against one population does not necessarily mean those parameters will be the best choice for *other* populations. This is one area in which pure theory, CE hydro simulations, and population models can work hand-in-hand with observational comparisons—the more closely the functional form of our parameterizations comes to representing the physics, the more reasonable it should be to expect that good parameter choices are broadly applicable.

For studying CEE for a specific population, in principle one could take a forward modeling approach from an immediate pre-CE population to a directly-related post-CE population, to try to minimize the confusion introduced by other theoretical uncertainties. For example, one might compare the population of pre-CE binaries containing a red giant and main-sequence star (or white dwarf) to the post-CE binaries containing a white dwarf and main-sequence star (or double white dwarfs). For example, Nie et al. (2012) use the population of LMC red giants showing ellipsoidal light-curve variation as their "initial" population. With large enough populations, this approach would be complementary to trying to infer the pre-CE properties of the post-CE binaries.

The literature contains many comparisons between population models and population data. Here we will illustrate this literature with Han et al. (2002, 2003), who studied the formation of hot subdwarf stars. Most hot subdwarfs are formed through removal of the hydrogen-rich envelopes of their progenitors, either through stable mass transfer or CEE, so the population can provide simultaneous constraints on CEE and the stability of Roche-lobe overflow. For the subset of the hot subdwarfs that ignited helium in degenerate stellar cores, the timing of the envelope removal needs to be rather fine-tuned. So, for that subset of the population, the assumption that $\alpha_{\rm CE}$ is a global constant may be more justified than in most population calculations. (This is not so for the hot subdwarfs that ignite helium in non-degenerate conditions, or for those formed by mergers of helium white dwarfs.) Han et al. (2003) used "full" stellar and binary evolution calculations, as opposed to semi-analytic fits, and calculated λ for their pre-CE structures. Not only did Han et al. (2003) consider variations in $\alpha_{\rm CE}$, but also included a parameter $\alpha_{\rm th}$, representing the fraction of the internal energy of the envelope (both thermal and recombination) which could be used in ejecting the envelope. Han et al. (2003) also considered variations in the stability of mass transfer from giant donors, $q_{\rm crit}$, the ratio of the donor mass to the accretor mass above which mass transfer is unstable. From by-eye comparisons to the data for hot subdwarfs, they concluded the best-fitting population corresponded to $q_{\rm crit} = 1.5$, and $\alpha_{\rm CE} = \alpha_{\rm th} = 0.75$, although they accepted that the fit was imperfect and that "the values of $\alpha_{\rm CE}$ and $\alpha_{\rm th}$ cannot yet be precisely determined."

We expect that improvements in near-future data on Galactic binary populations will provide excellent opportunities for comparison to population models. Simply checking whether our best theoretical understanding of the physics can explain the observed populations is valuable, as is understanding the expected observational consequences of small changes in the physical assumptions. Comparison between good population models and the data may well help to constrain the physics. Nonetheless, care should be taken to make the physics inside population models as close to our best understanding as is feasible.

10.2.3 Double-core Common-envelope Evolution

Double-core common-envelope ejection, i.e., common-envelope evolution in which the envelopes of both pre-CE stars are ejected, would leave behind two newly-exposed cores (see also Section 5.8). This form of CEE was proposed to help with formation of double-neutron-star binaries (see, e.g., Brown, 1995; Dewi et al., 2006), although the supposed problem it was designed to avoid (hyper-accretion converting the inspiraling neutron star into a black hole) is no longer considered problematic. The predicted importance of double-core CEE varies greatly among population models. An observational indication of the importance of double-core common-envelope ejection, or even a demonstration that it does really occur, would be valuable. However, we know of no observational evidence from the population of double-neutron-star binaries that indicates what fraction of them, if any, were formed following double-core common-envelope ejection.

At least one lower-mass post-CE binary indicates that double-core CEE occurs in nature. PG 1544488 contains two helium-rich hot subdwarfs with a period of roughly half a day (Ahmad 2004). Justham et al. (2011) argued that, if both components have ignited helium, double-core evolution naturally explains the properties of PG 1544488, and moreover that this channel favors progenitor structures that could explain why the surface abundances of the stars in this unusual system are helium-rich. This double-core interpretation was supported by Şener (2014); who confirmed that the components have similar masses and argued that observational evidence exists that at least one component has ignited helium.

Another system that deserves consideration as a potential post-double-core system is the close binary in the center of the bipolar planetary nebula Henize 2-428. Both components of that 4.2 hr binary are hot and have similar temperatures and luminosities, and they were initially taken as having near-identical masses (Santander-García et al. 2015; see also García-Berro et al. 2016). This is superficially as expected for a post-double-core system. A recent reanalysis by Reindl et al. confirms the similar temperatures and luminosities of the components, but finds masses that are less similar to each other than in the earlier works—$0.66 \pm 0.11\ M_{\odot}$ and $0.42 \pm 0.07\ M_{\odot}$. Those masses are still potentially consistent with formation via double-core common-envelope ejection, although Reindl et al. propose a scenario in which the lower-mass core is a helium white dwarf that was re-heated during the most recent interaction phase (in which case understanding and quantifying that reheating would itself be interesting). It is unclear to us whether analysis of the Henize 2-428 planetary nebula could realistically distinguish between a single-core and double-core CE scenario for this binary.

10.3 Post-common-envelope Planetary Nebulae

As discussed in Section 10.2, the existence of close binary stars in the centers of some planetary nebulae has provided good evidence to support the idea that common-envelope ejection actually happens, and the properties of those central binaries have been investigated for clues about CEE. Here we discuss what we might learn about CE physics from the nebulae around post-CE binary systems. Nebulae around inferred post-merger stars are discussed in Section 10.4. As elsewhere in this Chapter, we do not attempt to provide a full review.

Observations of planetary nebulae have the potential to give information about the fluid flows in ejected envelopes, e.g., the velocities and angular distributions of the ejected material. Those flows are related to the role of CEE in affecting the diversity of shapes of planetary nebulae. High-resolution observations may also be able to infer whether un-ejected envelope material remains close to the post-CE binary, e.g., whether there is a rotationally-supported circumbinary disk, a torus surrounding the binary, or even material falling back onto the central binary.

Explaining the diverse shapes of planetary nebulae is a complicated subject in which numerous processes may be implicated (for reviews see, e.g., Balick 2002; De Marco 2009). Unfortunately for CE physics, planetary nebulae are not shaped just by the outflows from CE events, but are also strongly affected by post-CE evolution (e.g., due to irradiation by the central object, or a combination of slow and fast winds), motivating us to find preplanetary nebulae that are as young as possible in our search for a pristine post-CE shape. Even single stars are not trivial, since magnetic fields and the interaction between the early slow-moving mass loss, from the time when the star was still on the asymptotic giant branch, and the later fast wind from the hot central star, can also shape planetary nebulae (see, e.g., Kwok et al. 1978; Kwok 1982; Kahn & West 1985). We stress that post-CE binary stars are not the only types of binary stars which might shape planetary nebulae; we have sometimes found the expected influence of post-CE binaries difficult to disentangle from other binaries when reading the literature. Here we only intend to consider planetary nebulae thought to have been produced following CE ejection.

Early investigations of the potential role of CEE in shaping planetary nebulae include Morris (1981); Soker & Livio (1989); and Bond & Livio (1990). If the densest mass ejections from CEE are in the orbital plane, with lower-density and faster material ejected in the polar directions, then it seems natural to broadly associate post-CE systems with bipolar nebula shapes, although this is far from a detailed explanation of the observations. Later Han et al. (1995) and Soker & Rappaport (2000) concluded, on the basis of their population models, that specifically the formation of bipolar planetary nebula is consistent with being dominated by systems in which binary interactions cause the bipolar morphology.

Observationally, bipolar morphology is common. Using a sample of 30 post-CE planetary nebulae, Miszalski et al. (2009b) found that approximately a third are clearly bipolar nebulae, and further that as many as two-thirds of their sample are plausibly bipolar. Hillwig et al. (2016) showed that, for all eight planetary nebulae for which the orientation of the close central binary orbit could be measured, the inclination of the binary axis aligned with the polar axis of the planetary nebulae. Hillwig et al. (2016) then concluded that this confirms the importance of common-envelope ejection in the shaping of planetary nebulae. Even systems that do not display bipolar morphology have been confirmed as bipolar based on outflow kinematics. For example, Abell 63, which had not previously been classified as a bipolar nebula, was subsequently determined to be so by Mitchell et al. (2007); who found an expansion velocity of $17 \pm 1\mathrm{km\,s^{-1}}$ in the orbital plane and 126 ± 23 km $\mathrm{s^{-1}}$ in the polar directions.

Clearly, if it is true that CE ejection produces bipolar flows, with slower ejection in the orbital plane and faster flows along the binary axis, then reproducing these velocity and density structures should be a goal of CE simulations. For the purposes of using planetary nebulae to constrain the physics of CEE, rather than using CEE as a mechanism to explain the morphology of some planetary nebulae, ideal observations are high-resolution ones of young planetary nebulae.

ALMA has enabled some wonderful high-resolution imaging of young planetary nebulae containing binary stars, such as the spectacular images of HD 101584, as shown in Figure 10.2 (Olofsson et al. 2019; following Olofsson et al. 2015). Olofsson et al. conclude that the kinetic energy in the HD 101584 ejecta is an order of magnitude larger than the energy that could have been released through binary inspiral and release of gravitational potential energy. However, they state the binary period in this system is 150–200 days, so if this is the post-CE binary that was responsible for the ejection of the envelope, then it is an unusually wide one.

In the case of another young preplanetary nebula, IRAS 16342-3814, we do not even know if it contains a binary, although the bipolar symmetry of the nebula is taken as suggestive. Murakawa & Izumiura (2012) inferred a mass of $\approx 1 M_\odot$ in an equatorial torus with an inner radius of several hundred AU, and Sahai et al. (2017) used ALMA to distinguish four components in the nebula material, with inferred ages between $\approx$100 and 450 yr. Tafoya et al. (2019) argue for the presence of a jet-like molecular outflow with a kinematic age of $\sim$70–100 yr, and that the jet displays a sinusoidal oscillation with an opening angle of $\approx 2^\circ$ and an oscillation period of 60–90 yr.

One of the youngest known planetary nebulae, IRAS 15103-5754, was observed with ALMA by Gómez et al. (2018). They find a circumstellar molecular torus with a radius of 1000AU, expanding at approximately 23 km $\mathrm{s^{-1}}$, which they estimate to have a mass of 0.4–$1.0 M_\odot$. In this case there is a second point source, consistent with a binary companion, but lying outside the torus—at a separation of $\gtrsim 1300$ au from its center—that is inconsistent with being the companion star in a close, post-CE binary. Gómez et al. (2018) nonetheless speculate that the nebula was formed through CE ejection, and so the second point source was the third star in a triple system, but we are not aware that the post-CE nature of the central source has been confirmed.

Above we have stated that it would be ideal to compare hydrodynamic simulations of common-envelope ejection to the structures of young preplanetary nebulae. Beautiful as these observed examples are, they would be more useful for comparison to CE models if we could be certain that they *are* post-CE nebulae. We repeat that, even where nebulae are shaped by a binary it is important to try to distinguish the *post-CE* binaries.

Current simulations of CE ejection do not self-consistently produce bipolar jets, or jet-like structures (with the exception of the not-yet-explained collimated outflows discussed in Section 9.3). Some simulations have added jet-shaped kinetic energy input "by hand" (see Section 4.1.5). It is unclear to us whether, by the time these

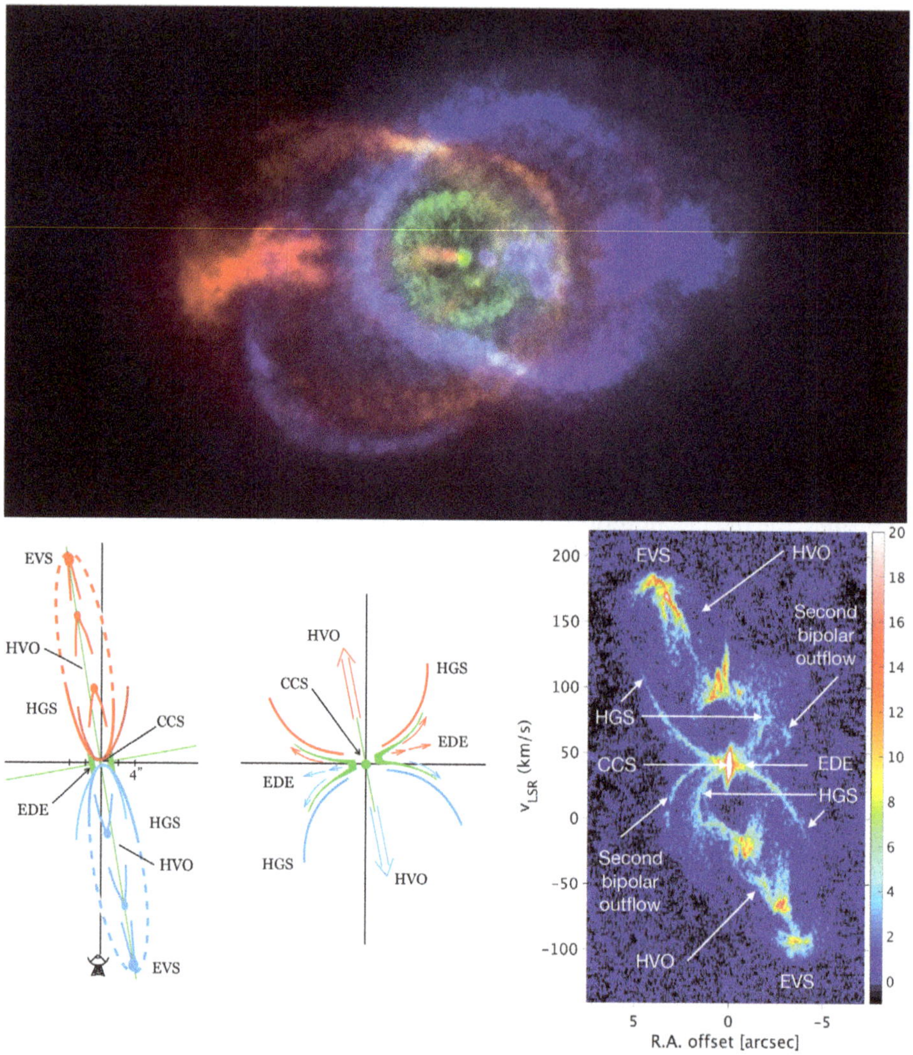

Figure 10.2. Top panel: False-color ALMA image of the environment of HD 101584. The colors represent speed: blue (red) for the gas moving fastest toward (away) from us. The stars in the binary are in the central dot within the green ring-like structure; green indicates material that is moving with the system velocity along the line of sight. (Credit: ALMA (ESO/NAOJ/NRAO), Olofsson et al Acknowledgement: Robert Cumming) Bottom panels: Different components of the circumstellar structure of HD 101584, from a perspective perpendicular to the line of sight. The left sketch shows the components with respect to the direction of the observer; the central panel sketches the inner regions. The components are labeled as the: central compact source (CCS); equatorial density enhancement (EDE); hourglass structure (HGS) which forms the inner part of two diametrically orientated bubbles; bipolar high-velocity outflow (HVO) with extreme-velocity spots (EVSs). An hourglass structure suggesting a tentative second bipolar outflow, with a velocity gradient opposite to that of the HVO, is also shown. The lower-right panel shows the CO(2–1) intensity at different velocities along the line of sight, for angles with respect to a position angle of 90°, with the different components indicated. The color scale indicates flux in mJy beam^{-1}. The bottom panels are from Olofsson et al. (2019), reproduced with permission from Astronomy & Astrophysics, © ESO.

structures are observed as preplanetary nebulae, isotropic fast winds from the central star would have operated long enough to accelerate gas to the observed speeds in the less-dense polar directions, or whether these fast bipolar outflows indicate a truly jet-like source.

Finally, related to the question of which planetary nebulae were shaped by binary interactions, several authors have studied the fraction of planetary nebulae that contain a (close) binary star (e.g., Miszalski et al. 2009a). This is clearly a directly relevant quantity when asking "how important are binary interactions for the population of observed planetary nebulae?" However, we encourage any readers who are tempted to infer the rate of common-envelope ejection relative to single-star envelope ejection from this fraction to be careful. This inference is sensitive to, e.g., whether post-CE and single-star planetary nebulae have different lifetimes. Iben & Tutukov (1989, 1993) and Hall et al. (2013) have investigated which exposed post-CE cores are expected to be able to ionize an ejected nebula so as to form a planetary nebula. Hall et al. (2013) conclude that degenerate cores less massive than $\approx 0.27_{\odot}$, if exposed by CE ejection on the first giant branch, would not produce a planetary nebula at all. Moreover, Hall et al. (2013) find the limiting core mass for producing a planetary nebula is sensitive to the criterion used to define how much of the envelope mass is ejected.

10.4 Presumed Post-merger Stars and Their Nebulae

A wide diversity of unusual stars, and stellar phenomena, have been interpreted as being a result of stellar mergers. Here we will confine ourselves to those mergers from failed common-envelope ejection, i.e., in which at least one of the stars is non-compact but has already left the main sequence.[5] Necessarily the identification of any unusual star as a post-merger object is a matter of interpretation, unless the merger has been directly observed. For example, it is tempting to interpret rapidly-rotating giant stars as post-merger objects (e.g., for FK Comae and V Hydrae, see Rappaport et al. 2009 and references therein, also Figure 1.3), but the plausibility of such a scenario does not constitute proof.

Perhaps the most famous individual example of a (likely) post-merger star is the progenitor of supernova 1987A. The star was a blue supergiant just prior to exploding, in conflict with all but extremely fine-tuned single-star evolution models. The properties of the supernova progenitor, including the unusual abundances in the envelope material, are naturally consistent with suitable post-merger models (see Podsiadlowski 1992, 2017; and references therein). Moreover, the triple-ring bipolar nebula seen around the site of the explosion can be compellingly explained as being a consequence of the merger; in this picture, equatorial mass shedding during the merger forms a torus that becomes the inner ring, and the outer two rings are naturally shaped by the interaction between that inner torus and the wind from the post-merger star (Morris & Podsiadlowski 2007, 2009). In this model, the initial

[5] Mergers of two main-sequence stars, or of two compact stars—or of a main-sequence star and a compact star—are nonetheless interesting subjects for study! Understanding those populations of mergers better would help with constraining, e.g., mass-transfer stability.

masses of the stars involved were $\approx 15 M_\odot$ and $\approx 5 M_\odot$, and the time delay between the merger and the explosion was approximately 20,000 yr—i.e., the donor star was a post-core-helium-burning extended giant star before the merger. If this model is correct, CEE between a red supergiant of $\approx 15 M_\odot$ and a main-sequence star of $\approx 5 M_\odot$ should fail to eject the majority of the envelope of the red supergiant.

Another massive star surrounded by a bipolar nebula with a distinct inner ring is the blue supergiant Sheridan 25 (Brandner et al. 1997a, 1997b). The fact that the center of the inner ring is somewhat offset from the bipolar axis, unlike in the remnant of supernova 1987A, may indicate that the ring was formed relatively promptly during the merger (Podsiadlowski et al. 2006; Morris & Podsiadlowski 2009).

The B[e] supergiant R4 is surrounded by a nitrogen-enriched cloverleaf-shaped nebula, suggesting that the ejected material was CNO-processed. Moreover, it is in a wide binary system with a superficially more evolved, but less luminous, star. Langer & Heger (1998) and Pasquali et al. (2000) resolved this discrepancy by explaining the observed B[e] supergiant in R4 as resulting from the merger of an inner binary soon after the more massive star left the main sequence. This naturally leads to the question of whether binary mergers are the dominant mechanism for the formation of B[e] supergiants (Langer & Heger 1998; Pasquali et al. 2000; Podsiadlowski et al. 2006).

Gallagher (1989), Iben (1999), and Podsiadlowski et al. (2006) discuss the question of whether a stellar merger was responsible for the Great Eruption of η Carinae, in which case the present-day star would be an early post-merger star, and the surrounding nebula would presumably contain an imprint of the earlier mass-loss history. Smith et al. (2018) combine information about the evolution of the outflow velocities before the outbursts with the outburst light-curve to try to construct a consistent history of the proposed merger.

Less extreme luminous blue variables (LBVs) than η Carinae may also include a significant contribution from post-merger stars. An animated debate has taken place in the literature about whether the observed LBVs are preferentially less tightly clustered than non-LBV stars with similar inferred masses, and have spatial distributions more consistent with stars of lower birth masses (for which see Smith 2019, and references therein). Relative isolation would be consistent with a stellar merger origin for many LBVs, because single stars with such high birth masses would be less likely to survive long enough to move to their observed separations. If they were main-sequence mergers, then such post-merger LBVs would not have passed through a "classical" CEE. However, numerous independent indications suggest that some supernova progenitors displayed LBV-like phenomenology soon before they exploded. Sufficiently massive main-sequence mergers may pass through LBV outbursts, but they would not be expected to explode as LBVs. However, direct LBV supernova progenitors might be explained by certain post-main-sequence mergers (specifically, more massive counterparts of the inferred R4 merger; see Justham et al. 2014; and references therein). An alternative CE-related explanation suggested for supernovae occurring in dense circumstellar environments is that these supernovae are triggered by the merger of a compact object into a stellar core at the end of a common-envelope phase (Chevalier 2012).

This takes us thematically back to stellar merger products as progenitors of core-collapse supernovae. As an extension of the merger model for supernova 1987A, the population of blue supergiant progenitors of core-collapse supernovae is an indirect indication that a population of suitable massive stellar merger remnants exists (Podsiadlowski et al. 1992). More speculatively, massive mergers for which *both* components were post-main-sequence at the time of the merger may explain some exotic massive-star supernovae (see Vigna-Gómez et al. 2019; and references therein); at present this is too far from the observations to be considered a robust conclusion. Nonetheless it seems plausible that future transient surveys, if combined with good models, could provide information to constrain the extragalactic population of massive stellar merger products from the supernova population.

10.5 Transients from CEE and Stellar Mergers

We have left perhaps the most topical and exciting observational constraints on CEE until (almost) the end of the book. Over the years, multiple transients have been proposed as potentially being direct consequences of mergers between non-compact stars. This was noted above for η Carinae (by, e.g., Gallagher 1989; Iben 1999 and Podsiadlowski et al. 2006). The spectacular outburst of V838 Monocerotis (Bond et al. 2003, and references therein) has also been interpreted via a stellar-merger model (e.g., Soker & Tylenda, 2003; Tylenda & Soker, 2006). In recent years there has not only been clear confirmation of a stellar merger origin for one of these events, but also a growing movement in favor of the idea that we should able to study common-envelope evolution using transients associated with envelope ejection.

An obvious first question is whether we should expect that common-envelope ejection typically leads to an optical outburst at all, in the sense of the system becoming more luminous across optical wavelengths as the envelope is ejected. To try to answer this, one might approximate the envelope structure of an ongoing CE event as that of a red giant star, close to the Hayashi line, and assume that the internal energy of the envelope is efficiently converted to mechanical energy as the envelope is ejected. If one further (incorrectly) assumes that the photosphere stays at a roughly constant mass coordinate as the envelope is ejected, then as the envelope departs, the photosphere expands and cools. For example, we might use $T \propto V^{1-\Gamma_1}$ for adiabatic expansion, where T and V are the temperature and volume of an ideal gas, and Γ_1 is the first adiabatic index. Then we can set $V \propto R^3$, where R is the radius of the photosphere, and simply assume for the bolometric luminosity of the photosphere that $L \propto R^2T^4$. Combining those assumptions, $L \propto R^2 \times (R^{3(1-\Gamma_1)})^4$. For $\Gamma_1 = 5/3$ this becomes $L \propto R^{-6}$ (or, for $\Gamma_1 = 4/3$, we have $L \propto R^{-2}$). Based on these assumptions there is no clear reason to expect an optical transient; instead the bolometric luminosity and temperature would both drop as the envelope departs. A red giant would disappear from the optical and move into the infrared, presumably becoming even redder as the cooling envelope forms dust. The exposed core would heat the ejected envelope from below, and it would eventually be revealed as the nebula dissipates.

Clearly the simple argument above makes several assumptions. Interested students reading this may wish to consider which of the above assumptions break down such that supernova explosions can look the way they do. Nonetheless, just because the supernova-driven ejection of the envelope of a red supergiant leads to an optical transient does not *guarantee* that the common envelope-driven ejection of the envelope of a red (super-)giant leads to a supernova-like light-curve.

10.5.1 V1309 Scorpii as the Cornerstone

The key observation that revealed that stellar mergers, at least, can produce bright optical events was OGLE photometry of V1309 Scorpii (Tylenda et al. 2011). The V1309 Sco outburst was discovered in early September 2008 (Nakano et al. 2008), and was soon proposed as being a near-twin of the V838 Mon transient (Mason et al. 2010). There were also broad similarities to a class of events that have generally come to be known as "Luminous Red Novae" (LRNe), including M85 OT2006-1 (Kulkarni et al. 2007; Rau et al. 2007), and the 1994 outburst of V4332 Sagittarii (Martini et al. 1999; Tylenda et al. 2005). The motivation for identifying these and similar events with stellar mergers was partly based on energetics, since the range of inferred energy output from these events was consistent with the range of energies that might be expected for stellar mergers. However, there was no direct proof.

Tylenda et al. (2011) supplied this proof for V1309 Sco, providing a vital, cornerstone discovery. Tylenda et al. (2011) showed that the pre-outburst light curve of the V1309 Sco binary matched that of a contact binary with orbital period $\approx$1.4 days, and moreover that the orbital period decreased measurably in the OGLE observations from August 2001 until February 2008, at which point the signs of binary periodicity were no longer present in the photometry. Moreover, in early 2008 the I-band luminosity was already starting to rise, well in advance of the main outburst in late August 2008. Tylenda et al. (2011) interpreted the disappearance of the orbital period signature as a sign of the system being obscured by matter flowing out from at least one of the outer Lagrangian points as the binary approached merger (see also Section 5.7 and Nandez et al. 2014). If February 2008 was the time of onset of outer Lagrangian point outflows, and if the fast rise toward the peak luminosity in late August 2008 took place soon after the time of the merger, then the time between the L2 outflows beginning and the merger was of order 100 orbital periods. Tylenda et al. (2011) further argued that the early pre-outburst rise in the system I-band luminosity is consistent with being due to emission from the matter in those outflows (see also, e.g., Pejcha et al. 2017).

10.5.2 Luminous Red Nova Light Curves

Ivanova et al. (2013) used the pre-merger properties of the V1309 Sco contact binary as initial conditions for hydrodynamic simulations of the merger event, which found that 0.03–0.08$M_{\odot}$ of matter was ejected during the merger (see also Nandez et al. 2014). Ivanova et al. (2013) combined this knowledge of the predicted amount of mass ejection with an analytic model for the emission from the ejected matter. This combination naturally explained the broad features from the outburst light curve.

Most notably, there was a plateau in the light curve lasting roughly 25 days. During the plateau in the V1309 Sco outburst light curve the inferred bolometric luminosity stayed approximately constant at $\approx 2 \times 10^4 L_\odot$. The inferred photospheric temperature and radius also stayed approximately constant during the plateau, roughly between 4000–5000 K and 150–300$R_\odot$ (Tylenda et al. 2011). This photospheric temperature is similar to that of the pre-outburst subgiant star inferred by Tylenda et al. (2011); which was ≈4500 K and ≈3.5 $R_\odot$ in radius.

Given that the matter ejected in the Ivanova et al. (2013) hydrodynamic model was from the outer envelope of the subgiant, readers may see that we must depart from the simple model of adiabatic cooling described at the start of this section. Assuming $T \propto R^{-2}$ for $\Gamma_1 = 5/3$, working backwards from 4500 K at 200$R_\odot$ gives, at 3.5$R_\odot$, a temperature of $\approx 1.5 \times 10^7$ K! This is a temperature not consistent with the outer envelope matter of a normal low-mass subgiant.

In supernovae, a way to overcome losses to adiabatic expansion is by heating the ejected material *after* it has expanded, as a consequence of radioactive decay. Here we cannot appeal to input from decay of radioactive elements. Instead heating may occur after the material has already expanded, thanks to input from hydrogen recombination[6] (as explored for tidal disruption events by Kasen & Ramirez-Ruiz 2010). For V1309 Sco, Ivanova et al. (2013) found that the amount of energy radiated away during the plateau is consistent with the amount of energy released by recombination of the matter ejected in their hydrodynamic simulations of the merger (on the other hand, the radiated energy was a factor of 5–60 less than the kinetic energy of the ejected material at infinity; (Nandez et al., 2014, see also below).

Hydrogen recombination has another effect on the ejected matter. There is a sharp decrease in the Rosseland mean opacity of the ejected matter after hydrogen has recombined. The location of the photosphere is set by hydrogen recombination in "Type IIP" supernovae, a class of core-collapse supernova that show a plateau in their light curves (see, e.g., Grassberg et al. 1971; Popov 1993).[7] If the location of the photosphere is set by hydrogen recombination, the temperature and size of the photosphere are self-regulated to be roughly constant even as the matter expands, leading to a phase in which the luminosity from the photosphere is approximately constant. So in those cases the photosphere of the ejecta is not fixed in mass coordinate; instead it moves inwards in mass coordinate as the density and temperature of the ejected matter evolve. The application of such a model to LRNe is consistent with the fact that LRNe all show a weeks-long plateau in their light curves, and moreover that the inferred photospheric expansion velocity during that phase is smaller than the expansion velocity inferred from spectra.

[6] In principle, helium recombination could also contribute.

[7] The "P" in Type IIP is for "plateau"—not to be confused with any "IIp" in which the "p" is intended to indicate "peculiar." The "II" indicates the presence of hydrogen in the spectrum (for which see Minkowski 1941; and note that the "I" and "II" classifications were intended to be *provisional*—in 1941).

Ivanova et al. (2013) adopted the Popov (1993) self-similar analytic model, which was derived for the case of Type IIP supernovae[8] and combined it with their own analytic parameterization for plausible properties of the ejected matter from stellar mergers and CE ejections. For these assumptions Ivanova et al. (2013) found that the predicted parameter space of optical transients from stellar mergers and common-envelope ejections naturally covered the parameter space of observed Luminous Red Novae. Later observed events have continued to fit that predicted parameter space (e.g., Dong et al. 2015; MacLeod et al. 2017; Blagorodnova et al. 2017), although—especially at the luminous end of the distribution of events—there continues to be uncertainty in knowing which events should be included as a probable consequence of stellar mergers or CE ejection (see, e.g., Pastorello et al. 2019). Howitt et al. (2020) published population predictions using variations of the Ivanova et al. (2013) model, and broadly reproduced the trends in the observed set of events.

However, this promising broad agreement between the models and observed events does not mean that all the problems are solved. Perhaps most importantly, the fact that the amount of energy radiated away during the plateau is consistent with the amount of energy released by recombination of the ejected matter for V1309 Sco does *not* mean this is true in all cases. The total amount of energy that is predicted to be radiated away during the plateau phase grows more steeply than linearly with the total ejected mass, such that the energy radiated away by the more luminous of the LRNe cannot be dominated by hydrogen recombination. (Recombination can still control the location of the photosphere during the light-curve plateau for those more energetic cases, as for Type IIP supernovae.)

Ivanova et al. (2013) speculated that shock heating might be significant for the energy budget of the transients for which recombination energy is insufficient (see also Metzger & Pejcha 2017). Indeed, even for V1309 Sco, the models of Nandez et al. (2014) found the excess kinetic energy of the ejected matter in their simulations was 5–60 times higher than was radiated away during the transient. If this kinetic energy can be appropriately thermalized, e.g., by internal collisions in the ejecta, then such kinetic energy could be converted into the light of the transient.

Such potential complications help to emphasize that the application of the Popov (1993) analytic supernova light-curve model is only a first step toward understanding even the population of LRNe that seem to fit the predictions that result from applying that model. Hydrodynamical models of ejections from stellar mergers and common-envelope ejections produce both spherical and aspherical ejecta, with multiple phases of ejection, and multiple ejection velocities (see Chapter 8). The systematic study of LRN light curves is still relatively underdeveloped compared to supernova modeling. No model has yet been published that self-consistently locates the effective photosphere in such a model, and so directly calculates a light curve. Indeed, no self-consistent light-curve calculation for any kind of common-envelope ejection has yet been published. A self-consistent simulation would include cooling

[8] They first re-derived the same expressions when starting with an envelope that is not dominated by radiation pressure, for which credit goes to J. Lombardi.

of the ejecta, and so cannot be done by post-processing of a purely hydrodynamical model.

However, we want to emphasize that because one merger event can lead to multiple ejections, multiple light-curve features could naturally result from one merger. For example, the Ivanova et al. (2013) hydrodynamic models for V1309 Sco produced multiple ejections. Ivanova et al. (2013) demonstrated that a preliminary light-curve model that combined emission from two such separate ejections improved their fit to the observed light curve of V1309 Sco. MacLeod et al. (2017) argue that the pre-plateau peak in the light curve of M31 LRN2015 is explained by photons diffusing out of an early ejection of $\sim 0.01 M_{\odot}$ of material. Using the same recombination-controlled photosphere model as Ivanova et al. (2013) they find the later plateau is consistent with an ejection of several tenths of a solar mass (see also Dong et al. 2015).

Even though these models are preliminary, with a great deal of room for improvement, the promising correspondence between the models and the observations offers hope that we can constrain stellar mergers and CE events by observing populations of transients. For example, for the population of events for which hydrogen recombination controls the location of the photosphere, this model suggests that matter is ejected while the hydrogen is still at least partially ionized. (We stress that, even *if* it turns out to be the case that—for some CE ejections—hydrogen recombination may not be helpful in full for ejecting the envelope, the *helium* recombination energy could still be useful—see also, e.g., Ivanova et al. 2015; Clayton et al. 2017.)

We also stress that the fact that an observed set of events—LRNe—fits the broad predictions of this model does not guarantee that all stellar mergers or CE ejections should be expected to look like this.

In particular, in recent years numerous infrared transients have been discovered (Kasliwal et al. 2017; Jencson et al. 2019). Given the fact that the envelopes of CE systems may be large and AGB-like (see, e.g., Clayton et al. 2017), it could be natural for some CE events to be cool and dusty, and so observed mainly in the infrared. Metzger & Pejcha (2017) also note that the early, pre-merger outflows from systems destined to merge are a natural site for dust formation, and so suggest that stellar mergers are natural candidates for explaining some of these infrared transients. Indeed, Jencson et al. (2019) argue that one of the nine infrared transients in their sample is most consistent with being a LRN. This raises complications related to the rates at which dust grains are formed. Follow-up observations of the optical LRN M31-LRN-2015 revealed that the system formed an optically-thick dust shell approximately 4 months after the peak of the optical transient, with further dust formation after 1.5 years (Blagorodnova et al. 2020).

10.6 Stars Undergoing a Common-envelope Phase

The fantastic pre-merger observations of the V1309 Sco outburst, and the potential of pre-outburst photometry for other stellar merger or CE transients, have amplified the hope that we may be able to watch common-envelope evolution in progress (see Section 10.5). The initial orbital period of V1309 Sco was only 1.4 days, which meant we could watch many orbital cycles. If we are fortunate enough to catch the onset of the plunge in a red-giant binary, and identify it in time to observe it carefully, then the longer orbital timescale might allow improved time resolution within each orbit. Even if all we can observe during the plunge is expansion of the donor star, rather than any quasi-periodic phenomena associated with the trajectory of the inspiral, such observations would be wonderful for checking that our simulations describe reality.

The potential onset of mass-transfer instability is one of the ways in which the Galactic microquasar SS 433 is an interesting source. King et al. (2000) & Podsiadlowski (2001) noted that the system properties are naturally consistent with the donor star being in the Hertzsprung gap, and the mass-transfer phase potentially ending due to a delayed dynamical instability. Blundell et al. (2001) saw evidence for equatorial outflows from SS 433, which they discuss as potentially being from a common-envelope already being built up around the binary. Losing SS 433 as a microquasar would be a sad outcome, but it would hopefully be illuminating for the study of common-envelope evolution (especially because a compact object would be involved).

However, we should not pin our hopes on catching a sample of Galactic dynamical-timescale CE episodes. Kochanek et al. 2014) calculate a Galactic rate for CE events of two per decade (and a combined Galactic rate for CE events and stellar mergers of approximately double that). If a significant fraction of CE events pass through a thermal-timescale slow-spiral-in phase, then the Galaxy would contain a population of examples that we might study. For a Galactic rate of CE events of one per decade, assuming that all CE events go through a slow spiral-in phase, and adopting a timescale of 100 years as an estimate for a representative duration of the slow spiral-in, then the Galaxy contains $\approx$10 examples. If that population exists, then we need to identify them.

The calculations of Clayton et al. (2017) may provide a useful guide to identifying this population (see also Section 7.4.1). These one-dimensional simulations of a slow-spiral-in phase found potentially observable pulsations in temperature and luminosity, examples of which are shown in Figure 10.3. These stars are large and cool. The period of the pulsations seen in Clayton et al. (2017) is 3–30 years. However, systematic transient surveys like the Zwicky Transient Facility (ZTF) are relatively new, and thus far a star undergoing similar pulsations might appear to be experiencing a monotonic change rather than a long-duration pulsation.

Observations of this phase may also be difficult in the cases for which there is mass ejection. A minority of pulsations lead to mass ejection from the envelope, where each mass ejection event can remove $\sim 0.1 M_{\odot}$ (Clayton et al. 2017). This ejection mechanism is one of the potential ways to remove the common envelope and end the CE phase (see Section 8.4). However, since there are previous episodes of mass ejection, dust formation in the earlier ejections may lead to high local extinction (or reddening) around the star. Nonetheless, we encourage photometric surveys with multi-year baselines to look for evidence of such pulsational behavior in cool, luminous giants.

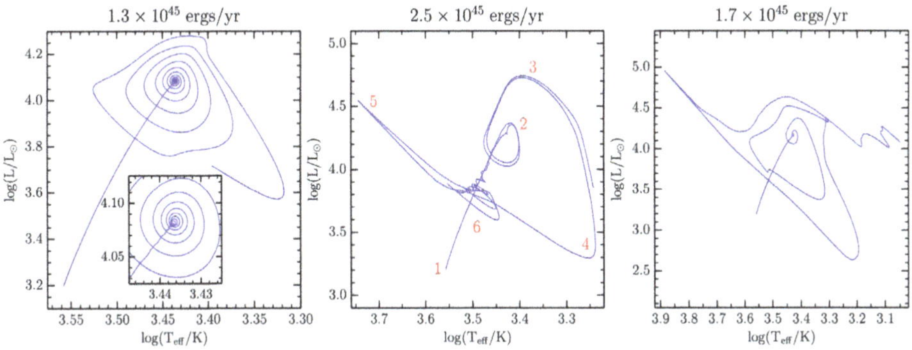

Figure 10.3. Tracks in the Hertzsprung-Russell diagram for one-dimensional models of the slow spiral-in phase, for three levels of heating (as labeled on the title of each panel). The tracks progress as shown in the middle panel, beginning at point 1 and progressing in numerical order to point 6, whereupon (if they repeat) they move back to point 2. Left: a case with no ejections, plotted until maximum radial amplitude is attained. Center: a case that exhibits a repeated cycle of strong envelope shocks but no ejections, plotted until after the first complete cycle. Right: a case that undergoes a large ejection; in this case the track follows the model photosphere of the ejected material. This figure is reproduced from Clayton et al. (2017).

Finally, in the first Chapter of this book we mentioned that one of the reasons why common-envelope evolution is of broader importance to astrophysics is that some post-common-envelope systems are thought to be detectable gravitational-wave sources. Here the circle is complete, since one of the hopes for future gravitational-wave detectors is that they may enable us to observe the ongoing inspiral of CE events (Holgado et al. 2018; Ginat et al. 2020).

References

Abell, G. O. 1966, ApJ, 144, 259

Ahmad, A., Jeffery, C. S., & Fullerton, A. W. 2004, A&A, 418, 275

Balick, B., & Frank, A. 2002, ARA&A, 40, 439

Blagorodnova, N., Karambelkar, V., Adams, S. M., et al. 2020, MNRAS, 496, 5503

Blagorodnova, N., Kotak, R., Polshaw, J., et al. 2017, ApJ, 834, 107

Blundell, K. M., Mioduszewski, A. J., Muxlow, T. W. B., Podsiadlowski, P., & Rupen, M. P. 2001, ApJL, 562, L79

Bond, H. E. 1976, PASP, 88, 192

Bond, H. E., Henden, A., Levay, Z. G., et al. 2003, Natur, 422, 405

Bond, H. E., Liller, W., & Mannery, E. J. 1978, ApJ, 223, 252

Bond, H. E., & Livio, M. 1990, ApJ, 355, 568

Brandner, W., Chu, Y.-H., Eisenhauer, F., Grebel, E. K., & Points, S. D. 1997a, ApJL, 489, L153

Brandner, W., Grebel, E. K., Chu, Y.-H., & Weis, K. 1997b, ApJL, 475, L45

Brown, G. E. 1995, ApJ, 440, 270

Burleigh, M. R., Steele, P. R., Dobbie, P. D., et al. 2011, in AIP Conf. Ser. 1331, Planetary Systems Beyond the Main Sequence, ed. S. Schuh, H. Drechsel, & U. Heber (New York: AIP), 262

Casewell, S. L., Braker, I. P., Parsons, S. G., et al. 2018, MNRAS, 476, 1405

Chevalier, R. A. 2012, ApJL, 752, L2
Clayton, M., Podsiadlowski, P., Ivanova, N., & Justham, S. 2017, MNRAS, 470, 1788
Davis, P. J., Kolb, U., & Knigge, C. 2012, MNRAS, 419, 287
Davis, P. J., Kolb, U., & Willems, B. 2010, MNRAS, 403, 179
de Kool, M., & Ritter, H. 1993, A&A, 267, 397
De Marco, O. 2009, PASP, 121, 316
De Marco, O., Passy, J.-C., Moe, M., et al. 2011, MNRAS, 411, 2277
Dewi, J. D. M., Podsiadlowski, P., & Sena, A. 2006, MNRAS, 368, 1742
Dhillon, V. S., & Marsh, T. R. 1995, MNRAS, 275, 89
Dong, S., Kochanek, C. S., Adams, S., & Prieto, J. L. 2015, ATel, 7173, 1
Drake, J. J., & Sarna, M. J. 2003, ApJL, 594, L55
Farihi, J., & Christopher, M. 2004, AJ, 128, 1868
Gallagher, J. S. 1989, in Astrophysics and Space Science Library Vol. 157, Proc. IAU Coll. 113, Physics of Luminous Blue Variables, ed. K. Davidson, A. F. J. Moffat, & H. J. G. L. M. Lamers (Dordrecht: Kluwer), 185
García-Berro, E., Soker, N., Althaus, L. R. G., Ribas, I., & Morales, J. C. 2016, NewA, 45, 7
Geier, S., Schaffenroth, V., Drechsel, H., et al. 2011, ApJL, 731, L22
Ginat, Y. B., Glanz, H., Perets, H. B., Grishin, E., & Desjacques, V. 2020, MNRAS, 493, 4861
Gómez, J. F., Niccolini, G., Suárez, O., et al. 2018, MNRAS, 480, 4991
Grassberg, E. K., Imshennik, V. S., & Nadyozhin, D. K. 1971, Ap&SS, 10, 28
Hall, P. D., Tout, C. A., Izzard, R. G., & Keller, D. 2013, MNRAS, 435, 2048
Han, Z., Podsiadlowski, P., & Eggleton, P. P. 1994, MNRAS, 270, 121
Han, Z., Podsiadlowski, P., & Eggleton, P. P. 1995, MNRAS, 272, 800
Han, Z., Podsiadlowski, P., Maxted, P. F. L., & Marsh, T. R. 2003, MNRAS, 341, 669
Han, Z., Podsiadlowski, P., Maxted, P. F. L., Marsh, T. R., & Ivanova, N. 2002, MNRAS, 336, 449
Hillwig, T. C., Jones, D., De Marco, O., et al. 2016, ApJ, 832, 125
Hoffleit, D. 1932, BHarO, 887, 9
Holgado, A. M., Ricker, P. M., & Huerta, E. A. 2018, ApJ, 857, 38
Howitt, G., Stevenson, S., Vigna-Gómez, A. R., et al. 2020, MNRAS, 492, 3229
Iaconi, R., & De Marco, O. 2019, MNRAS, 490, 2550
Iben, I. Jr 1999, in AIP Conf. Ser. 179, Eta Carinae at The Millennium, ed. J. A. Morse, R. M. Humphreys, & A. Damineli (San Francisco, CA: ASP), 367
Iben, I. Jr, & Tutukov, A. V. 1989, in IAU Symp. 131, Planetary Nebulae, ed. S. Torres-Peimbert (Dordrecht: Kluwer), 505
Iben, I. Jr, & Tutukov, A. V. 1993, ApJ, 418, 343
Ivanova, N., Justham, S., Avendano Nandez, J. L., & Lombardi, J. C. 2013, Sci, 339, 433
Ivanova, N., Justham, S., & Podsiadlowski, P. 2015, MNRAS, 447, 2181
Jencson, J. E., Kasliwal, M. M., Adams, S. M., et al. 2019, ApJ, 886, 40
Jones, D. 2018, arXiv:1806.08244
Jones, D., & Boffin, H. M. J. 2017, NatAs, 1, 0117
Justham, S., Podsiadlowski, P., & Han, Z. 2011, MNRAS, 410, 984
Justham, S., Podsiadlowski, P., & Vink, J. S. 2014, ApJ, 796, 121
Kahn, F. D., & West, K. A. 1985, MNRAS, 212, 837
Kasen, D., & Ramirez-Ruiz, E. 2010, ApJ, 714, 155
Kasliwal, M. M., Bally, J., Masci, F., et al. 2017, ApJ, 839, 88

King, A. R., Taam, R. E., & Begelman, M. C. 2000, ApJL, 530, L25
Kippenhahn, R. 1981, A&A, 102, 293
Kochanek, C. S., Adams, S. M., & Belczynski, K. 2014, MNRAS, 443, 1319
Kohoutek, L. 1967, BAICz, 18, 103
Kramer, M., Schneider, F. R. N., Ohlmann, S. T., et al. 2020, A&A, 642, A97
Kruse, E., & Agol, E. 2014, Sci, 344, 275
Kulkarni, S. R., Ofek, E. O., Rau, A., et al. 2007, Natur, 447, 458
Kwok, S. 1982, ApJ, 258, 280
Kwok, S., Purton, C. R., & Fitzgerald, P. M. 1978, ApJL, 219, L125
Langer, N., & Heger, A. 1998, in Astrophysics and Space Science Library, Vol. 233, B[e] Supergiants: What is Their Evolutionary Status?, ed. A. M. Hubert, & C. Jaschek (Dordrecht: Kluwer), 235
MacLeod, M., Macias, P., Ramirez-Ruiz, E., et al. 2017, ApJ, 835, 282
Mandel, I., Farr, W. M., & Gair, J. R. 2019, MNRAS, 486, 1086
Martini, P., Wagner, R. M., Tomaney, A., et al. 1999, AJ, 118, 1034
Mason, E., Diaz, M., Williams, R. E., Preston, G., & Bensby, T. 2010, A&A, 516, A108
Maxted, P. F. L., Burleigh, M. R., Marsh, T. R., & Bannister, N. P. 2002, MNRAS, 334, 833
Maxted, P. F. L., Napiwotzki, R., Dobbie, P. D., & Burleigh, M. R. 2006, Natur, 442, 543
Metzger, B. D., & Pejcha, O. 2017, MNRAS, 471, 3200
Miller, J. S., Krzeminski, W., & Priedhorsky, W. 1976, IAU Circ., 2974, 2
Minkowski, R. 1941, PASP, 53, 224
Miszalski, B., Acker, A., Moffat, A. F. J., Parker, Q. A., & Udalski, A. 2009a, A&A, 496, 813
Miszalski, B., Acker, A., Parker, Q. A., & Moffat, A. F. J. 2009b, A&A, 505, 249
Miszalski, B., Boffin, H. M. J., & Corradi, R. L. M. 2013, MNRAS, 428, L39
Mitchell, D. L., Pollacco, D., O'Brien, T. J., et al. 2007, MNRAS, 374, 1404
Morris, M. 1981, ApJ, 249, 572
Morris, T., & Podsiadlowski, P. 2007, Sci, 315, 1103
Morris, T., & Podsiadlowski, P. 2009, MNRAS, 399, 515
Murakawa, K., & Izumiura, H. 2012, A&A, 544, A58
Nakano, S., Nishiyama, K., Kabashima, F., et al. 2008, IAU Circ., 8972, 1
Nandez, J. L. A., Ivanova, N., & Lombardi, J. 2014, ApJ, 786, 39
Nelemans, G., & Tout, C. A. 2005, MNRAS, 356, 753
Nelemans, G., Verbunt, F., Yungelson, L. R., & Portegies Zwart, S. F. 2000, A&A, 360, 1011
Nie, J. D., Wood, P. R., & Nicholls, C. P. 2012, MNRAS, 423, 2764
Olofsson, H., Khouri, T., Maercker, M., et al. 2019, A&A, 623, A153
Olofsson, H., Vlemmings, W. H. T., Maercker, M., et al. 2015, A&A, 576, L15
Paczyński, B. 1970, AcA, 20, 47
Paczynski, B. 1976, in IAU Symp. 73, Structure and Evolution of Close Binary Systems, ed. P. Eggleton, S. Mitton, & J. Whelan (Dordrecht: Reidel), 75
Paczyński, B., & Ziółkowski, J. 1968, AcA, 18, 255
Pasquali, A., Nota, A., Langer, N., Schulte-Ladbeck, R. E., & Clampin, M. 2000, AJ, 119, 1352
Pastorello, A., Mason, E., Taubenberger, S., et al. 2019, A&A, 630, A75
Pejcha, O., Metzger, B. D., Tyles, J. G., & Tomida, K. 2017, ApJ, 850, 59
Podsiadlowski, P. 1992, PASP, 104, 717
Podsiadlowski, P. 2001, in ASP Conf. Ser. 229, Evolution of Binary and Multiple Star Systems, ed. P. Podsiadlowski, S. Rappaport, A. R. King, F. D'Antona, & L. Burderi (San Francisco, CA: ASP), 239

Podsiadlowski, P. 2017, in Handbook of Supernovae (Berlin: Springer), 635
Podsiadlowski, P., Joss, P. C., & Hsu, J. J. L. 1992, ApJ, 391, 246
Podsiadlowski, P., Morris, T. S., & Ivanova, N. 2006, in ASP Conf. Ser. 355, ed. M. Kraus, & A. S. Miroshnichenko (San Francisco, CA: ASP), 259
Popov, D. V. 1993, ApJ, 414, 712
Pretorius, M. L., Knigge, C., & Kolb, U. 2007, MNRAS, 374, 1495
Rappaport, S., Podsiadlowski, P., & Horev, I. 2009, ApJ, 698, 666
Rau, A., Kulkarni, S. R., Ofek, E. O., & Yan, L. 2007, ApJ, 659, 1536
Rebassa-Mansergas, A., Zorotovic, M., Schreiber, M. R., et al. 2012, MNRAS, 423, 320
Refsdal, S., & Weigert, A. 1970, A&A, 6, 426
Reindl, N., Schaffenroth, V., Miller Bertolami, M. M., et al. 2020, A&A, 638, A93
Sahai, R., Vlemmings, W. H. T., Gledhill, T., et al. 2017, ApJL, 835, L13
Santander-García, M., Rodríguez-Gil, P., Corradi, R. L. M., et al. 2015, Natur, 519, 63
Schaffenroth, V., Barlow, B. N., Drechsel, H., & Dunlap, B. H. 2015, A&A, 576, A123
Schaffenroth, V., Barlow, B. N., Geier, S., et al. 2019, A&A, 630, A80
Schaffenroth, V., Geier, S., Heber, U., et al. 2014, A&A, 564, A98
Şener, H. T., & Jeffery, C. S. 2014, MNRAS, 440, 2676
Smith, N. 2019, MNRAS, 489, 4378
Smith, N., Andrews, J. E., Rest, A., et al. 2018, MNRAS, 480, 1466
Soker, N., & Livio, M. 1989, ApJ, 339, 268
Soker, N., & Rappaport, S. 2000, ApJ, 538, 241
Soker, N., & Tylenda, R. 2003, ApJL, 582, L105
Tafoya, D., Orosz, G., Vlemmings, W. H. T., Sahai, R., & Pérez-Sánchez, A. F. 2019, A&A, 629, A8
Toonen, S., & Nelemans, G. 2013, A&A, 557, A87
Tylenda, R., Hajduk, M., Kamiński, T., et al. 2011, A&A, 528, A114
Tylenda, R., & Soker, N. 2006, A&A, 451, 223
Tylenda, R., Soker, N., & Szczerba, R. 2005, A&A, 441, 1099
van der Sluys, M. V., Verbunt, F., & Pols, O. R. 2006, A&A, 460, 209
Vigna-Gómez, A., Justham, S., Mandel, I., de Mink, S. E., & Podsiadlowski, P. 2019, ApJL, 876, L29
Webbink, R. F. 1979, in IAU Coll. 53, White Dwarfs and Variable Degenerate Stars, ed. H. M. van Horn, V. Weidemann, & M. P. Savedoff (Hamilton: Univ. Waikato), 426
Webbink, R. F. 2008, in Astrophysics and Space Science Library, Vol. 352, Astrophysics and Space Science Library in Short-Period Binary Stars: Observations, Analyses, and Results, ed. E. F. Milone, D. A. Leahy, & D. W. Hobill (Berlin: Springer), 233
Zorotovic, M., Schreiber, M. R., Gänsicke, B. T., & Nebot Gómez-Morán, A. 2010, A&A, 520, A86
Zorotovic, M., Schreiber, M. R., & Parsons, S. G. 2014, A&A, 568, L9

www.ingramcontent.com/pod-product-compliance
Ingram Content Group UK Ltd.
Pitfield, Milton Keynes, MK11 3LW, UK
UKHW051852100926
8556IPUK00005B/596

9 780750 315616